MONOGRAPHS AND RESEARCH NOTES IN MATHEMATICS

Published Titles

Application of Fuzzy Logic to Social Choice Theory, John N. Mordeson, Davender S. Malik and Terry D. Clark

Blow-up Patterns for Higher-Order: Nonlinear Parabolic, Hyperbolic Dispersion and Schrödinger Equations, Victor A. Galaktionov, Enzo L. Mitidieri, and Stanislav Pohozaev

Complex Analysis: Conformal Inequalities and the Bieberbach Conjecture, Prem K. Kythe

Computational Aspects of Polynomial Identities: Volume I, Kemer's Theorems, 2nd Edition Alexei Kanel-Belov, Yakov Karasik, and Louis Halle Rowen

Cremona Groups and Icosahedron, Ivan Cheltsov and Constantin Shramov

Diagram Genus, Generators, and Applications, Alexander Stoimenow

Difference Equations: Theory, Applications and Advanced Topics, Third Edition, Ronald E. Mickens

Dictionary of Inequalities, Second Edition, Peter Bullen

Iterative Optimization in Inverse Problems, Charles L. Byrne

Line Integral Methods for Conservative Problems, Luigi Brugnano and Felice Iavernaro

Lineability: The Search for Linearity in Mathematics, Richard M. Aron, Luis Bernal González, Daniel M. Pellegrino, and Juan B. Seoane Sepúlveda

Modeling and Inverse Problems in the Presence of Uncertainty, H. T. Banks, Shuhua Hu, and W. Clayton Thompson

Monomial Algebras, Second Edition, Rafael H. Villarreal

Nonlinear Functional Analysis in Banach Spaces and Banach Algebras: Fixed Point Theory Under Weak Topology for Nonlinear Operators and Block Operator Matrices with Applications, Aref Jeribi and Bilel Krichen

Partial Differential Equations with Variable Exponents: Variational Methods and Qualitative Analysis, Vicenţiu D. Rădulescu and Dušan D. Repovš

A Practical Guide to Geometric Regulation for Distributed Parameter Systems, Eugenio Aulisa and David Gilliam

Signal Processing: A Mathematical Approach, *Second Edition*, Charles L. Byrne

Sinusoids: Theory and Technological Applications, Prem K. Kythe

Special Integrals of Gradshteyn and Ryzhik: the Proofs – Volume I, Victor H. Moll

Special Integrals of Gradshteyn and Ryzhik: the Proofs – Volume II, Victor H. Moll

Forthcoming Titles

Actions and Invariants of Algebraic Groups, Second Edition, Walter Ferrer Santos and Alvaro Rittatore

Analytical Methods for Kolmogorov Equations, Second Edition, Luca Lorenzi

Geometric Modeling and Mesh Generation from Scanned Images, Yongjie Zhang

Groups, Designs, and Linear Algebra, Donald L. Kreher

Handbook of the Tutte Polynomial, Joanna Anthony Ellis-Monaghan and Iain Moffat

Microlocal Analysis on Rˆn and on NonCompact Manifolds, Sandro Coriasco

Practical Guide to Geometric Regulation for Distributed Parameter Systems, Eugenio Aulisa and David S. Gilliam

Reconstructions from the Data of Integrals, Victor Palamodov

Stochastic Cauchy Problems in Infinite Dimensions: Generalized and Regularized Solutions, Irina V. Melnikova and Alexei Filinkov

Symmetry and Quantum Mechanics, Scott Corry

MONOGRAPHS AND RESEARCH NOTES IN MATHEMATICS

Diagram Genus, Generators, and Applications

Alexander Stoimenow

GIST College, Gwangju Institute of Science and Technology

Gwangju, South Korea

CRC Press

Taylor & Francis Group

Boca Raton London New York

CRC Press is an imprint of the Taylor & Francis Group, an **informa** business

A CHAPMAN & HALL BOOK

CRC Press
Taylor & Francis Group
6000 Broken Sound Parkway NW, Suite 300
Boca Raton, FL 33487-2742

ISBN 13: 978-1-4987-3380-9 (hbk)

Visit the Taylor & Francis Web site at
http://www.taylorandfrancis.com

and the CRC Press Web site at
http://www.crcpress.com

Contents

Preface

Knot theory, which was begun in the late 19th century from the illusory vortex atoms and later put on a mathematical footing by Alexander, Reidemeister, Dehn, and others, has grown in recent years from a topological craft to a field where essentially all of mathematics can play a role. Knots are being studied in connection with number theory [Mr], variational analysis [DS], and other subjects our mathematical progenitors could hardly have dreamed of.

Knot theory received a significant impetus from the "polynomial boom" of the 1980s, which brought up worthwhile successors to the classical Alexander polynomial, as well as quantum and Vassiliev invariants, and much more. It is now readily accepted that many features of the newer polynomials are better understood in terms of combinatorial properties of knot diagrams. Yet even this topic has already grown so vast that a book (and my knowledge in general) would not be sufficient to cover it. Thus my intention in writing this book from the onset was not to attempt such an overall introduction. Rather, I focus on a particular piece of this area, developed mainly from my own previous work, and to present an extensive account on the study of the canonical genus: the genus of knot diagrams. A cornerstone for this entire study was the harmlessly appearing Theorem 2.11.8, which states that diagrams of a given canonical genus can be described by means of a finite number of patterns that I term "generators."

This book grew out of a manuscript written primarily for researchers in topology and combinatorics, and much like a journal article. Ultimately, it was too long to be published in such a format. I have made some effort in expanding the introduction and the background to make the manuscript more readable, in keeping with its goal of presenting new research. Very basic knowledge from university courses on algebra (polynomials, convex combinations), calculus (closure, limits) and combinatorics (graphs) should, in principle, be sufficient to grasp major concepts and their relations. I hope that, in this form, this work is accessible for graduate-level students. But let me make it clear that its intrinsic goal remains to be a research monograph and to supply proofs for a number of theorems stated (and also used elsewhere). Thus, undergraduates or the lay audience should be advised not to approach this as a textbook, and as I mentioned, I do not intend to give a general introduction to knot theory or even to combinatorial knot theory. Therefore, treatment of particular topics is not designed to be uniform; certain aspects will receive heavy emphasis, whereas others will only be mentioned marginally (if at all) and often disproportionally to (and without intention to diminish) their overall importance.

For more broadly suited general expositions, there are a number of other textbooks [Kf4, L, Ro, BZ]. Each one has its own viewpoint, but I consider Cromwell's recent work [Cr2] to be the closest in style to the topic of the present book. (In fact, Cromwell discusses therein the basic idea very briefly in Section 5.3.) Rather, my attitude in providing more background here is to make a larger circuit of readers interested in and possibly motivated to work on this particular subject. One of the things that fascinated me about knots was the multitude of problems that can be simply formulated, yet whose resolution was, and in many cases still remains, a long and painstaking ordeal. We will witness this in our efforts on several such problems. There are other problems we will barely be able to touch upon (and much more—enough for everyone to satisfy his/her own—and to dispute everyone else's—taste). For an almost completely different perspective of combinatorial knot theory, see, for example, the book [CDM].

This book can be briefly and roughly divided as follows.

1. (Chapters 1 and 2) Introduction and preliminaries. The topic is first introduced starting from—and leaning on—the very general exposition in [St22]. Then the content of the research is summarized. To prepare for a detailed discussion, I compile some background, informative in less detail than a textbook, but with enough references for further study.

2. (Chapter 3) Generator crossing estimates. We derive a new description of generators using Hirasawa's algorithm. This description leads to good estimates for the maximal number of crossings of generators. This part applies for general genus and is most extensively used outside the monograph.

3. (Chapter 4) Generators of genus 4. We show how the preceding description of generators can be used in practice to push the compilation of knot generators one step (=genus) further and complete their classification for genus 4. This part is visibly short but provides a key link between previous and subsequent material.

4. (Chapters 5 to 9) Applications of the genus 4 classification. Here we follow the line of thought that, with the list of generators at hand, one can design efficient algorithms to address in the special case of low genus several, including long-standing, problems.

   We establish non-triviality of the skein (HOMFLY-PT) polynomial on k-almost positive knots for $k \leq 4$, and of the Jones polynomial for $k \leq 3$. For $k \leq 4$, we classify the occurring achiral knots and prove a trivializability result for k-almost positive unknot diagrams. This also yields estimates for the number of unknotting Reidemeister moves. We describe the positive knots of signature (up to) 4.

   Using a study of the skein polynomial, we prove the exactness of the Morton–Williams–Franks braid index inequality and the existence of a minimal string Bennequin surface for alternating knots up to genus 4. We also prove for such knots conjectures of Hoste and Fox about the roots and coefficients of the Alexander polynomial.

5. (Chapter 10) Perspective. The above summarized content only partially covers the applications of the generator idea. To put the work here into a somewhat broader context, I added a brief final chapter describing previous and outlining subsequent related work.

As in a research monograph, I cannot present all proof details at a level as elementary as I wrote for parts of the introduction. It is a painful experience that writing mathematics in a readable (and even more so, pleasant) way tends to be more verbose than one likes. I also made no attempt to significantly alter the content of research from the original version. Of course, there is less space pressure in books than in journals, and books make it possible to include a different (e.g., anecdotal) style of material, which was an opportunity I was happy to take in a few places. Nevertheless, I feel that the scope even in books should not be extended without a clear objective. I thus hope that in the present form it is a self-contained and worthwhile exposition.

Gwangju, December 2015

The author can be contacted at `stoimeno@stoimenov.net`. Some computational material related to the book is available on the author's website `http://stoimenov.net/stoimeno/homepage/`.

Acknowledgments

I wish to thank L. Kauffman for inviting me to Waterloo in the fall of 2003 and for letting me present there, in a talk, some of the work concerning this book. The material in Chapter 5 was motivated by some discussion with T. Tsukamoto, and the one in §5.6 by discussion with A. Holmsen. Furthermore, S. Rankin, M. Thistlethwaite, and K. Fukuda provided some technical help for using their programs. Ellis Lee has offered valuable feedback about language editing.

Part of this research was supported by Postdoc grant P04300 of the Japan Society for the Promotion of Science (JSPS) at the University of Tokyo (I wish to thank my host T. Kohno for his support), the 21st Century COE program at RIMS (Kyoto, Japan), and the BK21 Project at KAIST (Daejeon, Korea).

A final thanks must be given to the staff at T & F, and in particular to Bob Stern. It was only through his encouragement that the initially vague project of making a book became reality.

List of Figures

List of Tables

Symbol Description

Symbols (except those with very local meaning in proofs) are listed below in the order they are introduced (with occasional reference to the section/chapter; if missing, see the reference of the previous symbols).

D	A diagram of a knot or link (§2.1)
$n(.)$	The number of components of a diagram or knot/link
#	Connected sum of links or diagrams (§2.1)
$c(.)$	The crossing number of a diagram of a knot/link (§2.2)
$c_\pm$	The number of positive/negative crossings of a diagram
$w(D)$	The writhe of a diagram
!	Mirror image of a diagram or link (§2.2)
$g(.)$	The canonical genus of a diagram D or the Seifert genus of a knot or link (§2.4)
$\chi(.)$	Euler characteristic
$s(D)$	The number of Seifert circles
$\tilde{g}(.)$	The canonical genus of a knot or link (§2.4)
$s_-(D)$	The number of negative Seifert circles
$\operatorname{ind}(s)$	The index of a Seifert circle s
G	A graph (§2.5)
$V(G)$ $E(G)$	Vertices and edges of a graph (§2.5)
$\operatorname{star} v$	Star of a vertex in a graph
G^*	Dual graph
B_n	Braid groups (§2.7)
σ_i	Standard Artin braid generators
$b(L)$	Braid index (§2.7)
min deg	Minimal degree (§2.8)
max deg	Maximal degree (§2.8)
span	Span of a polynomial (§2.8)
$[X]_d$	Polynomial coefficient (§2.8)
P	Skein (HOMFLY-PT) polynomial
V	Jones polynomial
Δ	Alexander polynomial
∇	Conway polynomial
F	Kauffman polynomial
A(D)	The (bracket) A- state of D
$a_\pm(D)$	Thistlethwaite's bounds for the number of positive/negative crossings (§2.8)
$\operatorname{mwf}(L)$	The Morton–Williams–Franks bound for the braid index (§2.9)

G/v	Star-contracted graph
$G\setminus v$	Star-deleted graph
$*$	Block (Murasugi) sum of diagrams or join of graphs (§2.9)
$\operatorname{ind},\operatorname{ind}_\pm$	Index, positive/negative index of a graph (§2.9)
$\Gamma(D)$	The Seifert graph of a diagram D
$\operatorname{mpb}(D)$	The Murasugi–Przytycki bound for the braid index from a diagram
σ	The signature (§2.10)
$p\cap q$	Linked crossings p,q in a diagram (§2.11)
$p\simeq p$ $p\,\tilde{*}\,q$ $p\sim q$	Equivalences of crossings
$t(D)$	Number of $\sim$-equivalence classes of a diagram
$t_i(D)$	Size of i-th $\sim$-equivalence classes of a diagram
$p(D,i,j)$	Numbering of crossings in a diagram w.r.t. $\sim$-equivalence classes
$\langle D\rangle$	The series of a diagram
G'	Reduced bisection of a planar 3-valent graph G (Chapter 4)
$O(.)$	Big O-notation (Chapter 4)
$\operatorname{len} b$	Length of a bridge in a diagram (§5.2)
α	Vector of sizes of negative $\sim$-equivalence classes (§5.2)
D_*,D_+	Basic test diagrams (§5.2,5.3)
$\deg C$	(Flyping) degree of a $\sim$-equivalence class (§5.3)
$r(D)$	Number of trivializing Reidemeister moves (§5.6)
$G/_e v$	Corrected star-contraction in graph (to correspond to the Murasugi–Przytycki diagram move; §7.3)
$G//_e v$	Restricted star-contraction for Bennequin surfaces (§8.2)
ind_b	Restricted index for Bennequin surfaces (§8.2)
$\Delta_{[i]}$	i-th coefficient of the Alexander polynomial (§9.1)
∇_i	Sign-adjusted i-th coefficient of the Conway polynomial (§9.3)
$tb(\mathcal{K})$ $\mu(\mathcal{K})$	Thurston–Bennequin number and Maslov (rotation) index of a Legendrian knot (§10.1)
$S_{G,O}$ $L_{G,O}$	Thickening of a marked 3-valent graph, and its boundary (§10.3)
vol	Hyperbolic volume of a link or a graph (§10.5)
$g_f(K)$	Free genus of a knot
L_G	Limit link of a graph
W_N	The sl_N polynomial of a 3-valent graph (§10.6)

Chapter 1

Introduction

1.1 The beginning of knot theory

In the late 19th century, W. Thomson ("Lord Kelvin"), P. G. Tait and J. Maxwell initiated the vortex-atom theory in an attempt to explain the structure of the universe. They believed that a super-substance – ether – makes up all matter and that atoms are knotted tubes of ether. Knotting is thereby understood as tying a piece of rope and then identifying both of its ends so that the tying cannot be any more undone.

Thus, in the realm of constructing a periodic table of elements, Tait began cataloging the simplest *knots*. He depicted knots (as we still do today) by *diagrams*, consisting of a (smooth) plane curve with transverse self-intersections, or *crossings*. At each crossing, one of the two strands passes over the other. The above term "simplest" refers to the number of crossings of the diagram.

We can define the *crossing number* of a knot as the minimal crossing number of all its diagrams and can say that Tait sought the list of knots with a given (small) crossing number.

There is not much difficulty in creating arbitrarily complicated diagrams of the same knot from a given one; thus, the non-trivial question is whether and to what extent one can *reduce* the crossing number. This question is very difficult in general, being one of the cornerstones around which all of knot theory was built. However, in certain situations, the answer is easy. In diagrams of the form 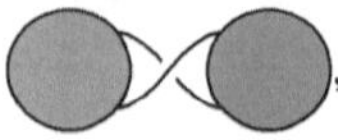, the depicted crossing can be easily eliminated (see Figure 2.1). We call such diagrams *reducible*, and all others *reduced*.

Tait's list sought to present each knot by exactly one diagram. This requires that knots from different diagrams should be inequivalent in the sense that one cannot turn a (closed) piece of rope which is knotted one way into a piece knotted another way without cutting the rope. The simplest knots are shown in Figure 1.1. The left-most one, of crossing number 0, is the *trivial* knot or *unknot*. It has some special importance, much like the unit element in a group.

unknot right-hand trefoil figure-8 knot

FIGURE 1.1: The simplest knots

Tait's first table went to 7 crossings. His accomplishment entitles him to be called the first knot theorist. However, Tait worked mainly by intuition. He had at his time no rigorous tools to show knots inequivalent. Little, Kirkman, and later Conway [Co] and others took over and continued his work. The efforts to compile knot tables have gone on for more than a century after those early pioneers even though Tait's vortex-atom theory has long been dismissed. The last manually compiled tables were apparently completed in the 1980's for 11 crossing knots [Co, Ca, Pe2]. In the modern computer age, tables have reached knots of 17 crossings, with millions of entries. An account on knot tabulation is given—with emphasis on its history—in [S], and from a more contemporary point of view in [HTW, H]. See also §2.3 below.

Right from the start of knot tabulation, a suggestive property attracted some attention. A diagram is called *alternating* if the curve passes crossings interchangingly over-under like . A knot is alternating if it has such a diagram. It seems to remain unclear whether Tait was convinced certain properties hold for all knots or just for alternating knots. The knots in Figure 1.1 are alternating. In fact, so are all knots up to 7 crossings, cataloged by Tait, and at least a large portion of the slightly more complicated ones he saw during his lifetime from his successors' work (even though it is now known that alternation is a rare property for generic crossing numbers [Th3]). Thus, certainly Tait was guided by evidence from alternating diagrams.

This evidence would prove misleading in some cases (see the end of Section 2.8.3), but two of his main conjectures stood for a long time.

Conjecture 1.1.1 (Tait's Crossing Conjecture) A reduced alternating diagram has minimal crossing number (for the knot it represents).

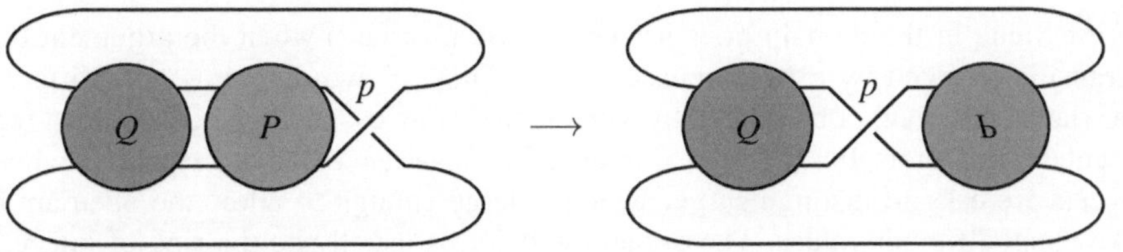

FIGURE 1.2: A flype near the crossing p.

Conjecture 1.1.2 (Tait's Flyping Conjecture) Alternating diagrams of the same knot are related by a sequence of *flypes* (shown in Figure 1.2).

1.2 Reidemeister moves and invariants

A few decades after Tait, the work of Alexander, Reidemeister and others began to put knot theory onto a mathematical footing. Reidemeister showed that three types of local moves suffice to interrelate all diagrams of the same knot.

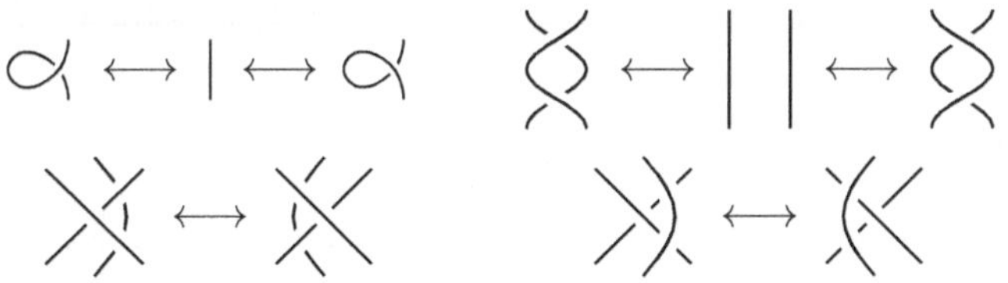

These moves are *local* in the sense that they only alter a fragment of the diagrams, as shown above. The concept of local changes of diagrams (not necessarily preserving the underlying knot or link) is a central theme in knot theory and even more so for this book. Thus, from now on we will understand that whenever a local move is depicted to change between two diagrams, only the altered part will be depicted while the rest of the diagrams will be assumed to be equal.

To distinguish two knots thus translates into the question how to prove that two diagrams (of these knots) are not connected by a sequence of Reidemeister moves. The problem is that such a sequence may be arbitrarily long and pass through very complicated intermediate diagrams. (There is a fundamental issue that one can to some extent control these sequences [HL], and how to do so. We will treat this problem in §5.6.)

Excluding a sequence of Reidemeister moves is done with the help of an *invariant*, that is, a map

$$\{ \text{ knot diagrams } \} \to \text{ some domain},$$

whose value in the domain does not change (is invariant) when the argument (diagram) is changed by a Reidemeister move. Thus, if two diagrams give different invariants, they cannot be related by Reidemeister moves, and hence, they must represent distinct knots. For the value domain of the invariant one thus seeks a set whose objects are easy to distinguish yet which is large enough to allow the invariant to take many different values. The domain will for us usually be the ring of integers, or of (Laurent) polynomials with integer coefficients. Clearly there are many different polynomials, and one can more easily compare coefficients than wondering about a long sequence of Reidemeister moves. Another obvious desirable feature of an invariant is that we can easily evaluate it from a diagram.

Alexander's contribution was to construct precisely such an invariant. Using his methods, Tait's knot lists would be *proven* correct. The *Alexander polynomial* [Al] remained—and still remains—a main theme in knot theory for decades to come. Let us observe, though, that we actually already came across one other knot invariant (here with values in $\mathbb{N}$), the crossing number. It is an invariant directly by definition, being defined on the whole Reidemeister move equivalence class of diagrams. However, this definition makes it difficult to evaluate from a diagram; moreover, many knots have equal crossing numbers. In contrast, the Alexander's polynomial can be calculated readily by several procedures (see §2.8.2) and easily distinguishes most knots.

1.3 Combinatorial knot theory

The Alexander polynomial was, from its very beginning, connected to topological features of knots. This topological viewpoint remained a *status quo* in knot theory research for decades, until in the 1980's a new chapter was opened by V. Jones with the discovery of a successor to Alexander's polynomial. The developments the *Jones polynomial* V [J] has sparked in the 30 years since its appearance are impossible to even vaguely sketch in completeness. Immediately after the Jones polynomial, the skein (HOMFLY-PT), BLMH, Kauffman, and other polynomial invariants appeared, and later Vassiliev [BN3] and quantum invariants. A good related (though still partial, and now no longer very recent) account was given by Birman [Bi].

One of the first achievements of the Jones polynomial, though, was the solution of Tait's conjectures for alternating knots. The Crossing conjecture was proved by Kauffman [Kf2], Murasugi [Mu, Mu2], and Thistlethwaite [Th4] (see Theorem 2.8.5 below). Kauffman's proof uses a calculation procedure for V called *state model.* This state model also gives a very elementary proof that the Jones polynomial is an invariant. (Kauffman had previously developed a similar model for the Alexander polynomial, too [Kf3].) Tait's Flyping conjecture was resolved a few years later by Menasco and Thistlethwaite [MT] (Theorem 2.6.2 below), mostly by using geometric techniques, though again with some (now subordinate) appearance of the Jones polynomial.

The Jones polynomial and its successors are different from the Alexander polynomial in many ways. One manifestation of this difference is the slow progress toward an important long-standing question of Jones. (We will return to this problem in §5.5.)

Question 1.3.1 Does the Jones (or skein, etc.) polynomial detect the unknot? That is, if a knot has polynomial 1, is it the unknot?

For the Alexander polynomial, our topological understanding has long answered the question (negatively). But attempts to gain topologically more understanding of the newer polynomials failed to a large extent, and thus mathematicians went on looking for other ways. One of these ways was the development of combinatorial knot theory, which is the study of combinatorial properties of knot and link diagrams, a theme to dominate this book. I will focus on a particular piece of this area, developed mainly in my own previous work, with an extensive study of canonical genus. Below I am going to introduce this subject and give an overview of the work in this monograph.

1.4 Genera of knots

Introducing the *genus* $g(K)$ of a knot K, Seifert [Se] gave a construction of compact oriented surfaces in 3-space bounding the knot (*Seifert surface*) by an algorithm starting with some diagram of the knot (see [Ad, §4.3] or [Ro]). The surface given by this algorithm is called *canonical.* A natural problem is when the diagram is *genus-minimizing*, or *of minimal genus*, that is, its canonical Seifert surface has minimal genus among all Seifert surfaces of the knot. This problem has been studied over a long period. First, the minimal genus property was shown for alternating diagrams, independently by Crowell [Cw] and Murasugi [Mu2]. Their proof is algebraical, using the Alexander polynomial Δ [Al] and the inequality $\max\deg\Delta \leq g$ (which thus they proved to be exact for alternating knots). Later Gabai [Ga] developed a geometric method using foliations and showed that this method is likewise successful for alternating diagrams. Murasugi [Mu3] introduced the operation *-product (or Murasugi sum). It was shown to behave naturally w.r.t. the Alexander polynomial by himself, and later by Gabai [Ga2, Ga3] in the geometrical context. These results imply the extension of the minimal genus property of alternating diagrams to the homogeneous diagrams of [Cr]. Another important subclass of homogeneous diagrams and links are the positive diagrams and links. Such links have been considered (in general or in special cases) independently before. (See e.g., [CG, CM, FW, O, Ru, Tr, Yo, Zu].) The minimal genus property for such diagrams follows from yet a different source, the work of Bennequin [Be] on contact structures. His inequality (theorem 3 in that paper; stated below as Theorem 2.4.2) in fact allows estimating the difference between the genus of the diagram and the genus of the knot in terms of the number of positive or negative crossings.

1.5 Overview of results

1.5.1 General generator estimates

The prospect of applications led to the treatment of canonical surfaces and to establishing in its own right the *canonical genus*, which is the minimal genus of all such surfaces for a given knot or link. In this monograph, we will continue the study of canonical Seifert surfaces from the combinatorial point of view. This study was initiated by me in [St4], and independently by Brittenham [Br], and set forth in [St2], and later in [STV, SV]. (Some explanation of this work is also given in section 5.3 of Cromwell's recent book [Cr2].) Using the theoretical insight gained there, we had efficient methods to complete the classification up to genus 3, in terms of the list of "generators." In [SV] the relation to certain algebraic objects named Wicks forms was discussed and applied to the enumeration of alternating knots by genus (see §10.3). We gave an extensive list of other applications in [St2] and later, among others, in [St18, St12].

The previous method of [SV] does not apply well for several components, and thus we develop an alternative approach, which is based on the special diagram algorithm first found by Hirasawa [Hi2] (and rediscovered a little later independently in [St9, §7]). We will work out inequalities for the crossing number and number of $\sim$-equivalence classes of $\bar{t}'_2$-irreducible link diagrams (generators). So far we carried this out only for knots. The present approach allows us to improve what we know in the knot case and extend it to links.

1.5.2 Compilation of knot generators of genus 4

Then we treat the description and applications of the compilation of knot generators of genus 4. As has already become apparent in the preceding generator compilations for genus 2 and 3, and then also from the result of [SV], the growth of the number of generators is enormous. In order to push the task back within the limits of (reasonable) computability, we need important new theoretical knowledge. This insight relies heavily on the special diagram algorithm of Hirasawa and the previous careful analysis of this procedure which we carry out.

1.5.3 Unknot diagrams

The first application addresses the identification of the unknot. This a basic problem in knot theory. There are general methods using braid foliations [BM2] and Haken theory [HL]. However, these methods are difficult to use in practice. So one is interested in applicable criteria, at least for special classes of diagrams.

It is known, for example, from Crowell [Cw] and Murasugi [Mu2] that in alternating unknot diagrams, all crossings are nugatory. For positive diagrams the same phenomenon can be observed either using the Alexander polynomial (as it applies for alternating diagrams; see e.g., [Cr]), or as an application of the Bennequin inequality (Theorem 2.4.2 below) in its extended form based on the Vogel algorithm [Vo] (see [St]). For almost alternating diagrams a trivializability result using flypes and "tongue moves" (see Figure 2.8) was proved recently by Tsukamoto [Ts], confirming a previous conjecture of Adams. Here we prove several related results, *inter alia*, the following:

Theorem 1.5.1 For $k \leq 4$, all k-almost positive unknot diagrams are trivializable by crossing number reducing wave moves and factor slides.

A wave move is shown in Figure 2.7, and a factor slide in Figure 2.2 (b).

For k-almost positive diagrams the description in the case $k = 1$ (unknotted twist knot diagrams and possible nugatory crossings) was written in [St3]. It was, however, previously known to Przytycki, who observed it as a consequence of Taniyama's result [Ta]. Their joint work had been announced long ago, but the full paper was not finished until very recently [PT]. This description includes and concretifies Hirasawa's result [Hi] on almost special alternating diagrams. For 2-almost positive diagrams (in particular 2-almost special alternating diagrams) the result is given in [St2]. We now settle the cases $k \leq 4$ in the stated way.

This theorem is also quite analogous to the result[1] in [NO] for 3-bridge diagrams and for arborescent (or Conway-algebraic) diagrams, which is a consequence of Bonahon–Siebenmann's classification of arborescent links. Their manuscript was notoriously difficult to find but is now available on the web [BS]. A partial published account on their work, which covers the treatment of the unknot, is given in [FG].

A tongue move is a special type of wave move, preserving the property "almost alternating." From this point of view, we can extend Tsukamoto's (and Hirasawa's) result for $k = 1$ in special diagrams to $k \leq 5$ (see Corollary 5.4.2 and Theorem 5.5.4). In fact, this extension was one of the motivations for our interest in this type of

[1] I was informed, though, of a possible gap in its proof.

problem. We can also apply our work to give polynomial estimates for the number of Reidemeister moves needed for unknotting, which has recently been studied by various authors [HL, HN, Hy] (see Proposition 5.6.1).

From a different perspective, a factor slide is a (very special) type of preserving wave move. Then Theorem 1.5.1 confirms in our case, in a stronger form, a conjecture (Conjecture 5.4.4) which states that crossing number preserving and reducing wave moves in combination suffice to trivialize unknot diagrams. Example 5.4.1 shows that our stronger statement is not true in general, and the assumption $k \leq 4$ in the theorem cannot be improved.

1.5.4 Non-triviality of the link polynomials

The work on Theorem 1.5.1 allows us to address two related problems which have been given some attention in the literature. The following result relates to the non-triviality problem of link polynomials (Question 1.3.1).

Theorem 1.5.2 Any non-trivial k-almost positive knot has non-trivial skein polynomial P for $k \leq 4$ and non-trivial Jones polynomial V for $k \leq 3$.

So far, only the cases $k = 0, 1$ were known (see [St6]). For P, the result follows from the proof of Theorem 1.5.1 almost directly. For V, we combine the skein and Kauffman bracket properties of the Jones polynomial in [BMo, St6, St8] with some arguments about the signature and Gabai's geometric work [Ga2, Ga3]. (They allow us to reduce the necessary computations to a minimum.)

The proof of Theorem 1.5.1 can be applied in some form not only to the trivial but more generally to an amphicheiral (achiral) knot (see Corollary 5.7.3). We can then obtain the classification of these knots.

Theorem 1.5.3 The k-almost positive achiral knots for $k \leq 4$ are those of $2k$ crossings, that is, 4_1 for $k = 2$; 6_3 and $3_1\#!3_1$ for $k = 3$; and 8_3, 8_9, 8_{12}, 8_{17} 8_{18}, and $4_1\#4_1$ for $k = 4$.

This property was known for $k \leq 2$ by [PT]; prior to the latter's completion, written account was given in [St6] for $k \leq 1$ and [St2] for $k = 2$. Theorem 1.5.3 adds the cases $k = 3, 4$. It was obtained in [St6] for alternating prime knots (independently provable using the work in [Th]), and checked there also for prime knots of ≤ 16 crossings. The proof of Theorem 1.5.3 will consist in extending the arguments for Theorem 1.5.1 and simplifying diagrams of such knots to these checked low crossing cases (see Corollary 5.7.3).

1.5.5 Signature of positive knots

A final application concerning positivity is the determination of the positive knots of signature 4 (Theorem 6.2). Again the result for signature 2 (namely, that these are precisely the positive knots of genus 1) is a consequence of Taniyama's work [Ta]. Our theorem is the first non-trivial explicit step beyond Taniyama toward the general case of the conjecture that positive knots of only finitely many genera have a given signature; see [St15, St12].

1.5.6 Braid index and Bennequin surfaces

The later sections of this monograph extend the applications of generators and regularization to alternating knots. In Chapter 7, we use a regularization of the skein polynomial and the work of Murasugi–Przytycki [MP] to show

Theorem 1.5.4 The Morton–Williams–Franks braid index inequality MWF is exact on alternating knots of genus up to 4.

We will also confirm a conjecture of Murasugi–Przytycki (Conjecture 2.9.5) for such knots. Theorem 1.5.4 complements the results in [Mu] for fibered and 2-bridge knots; it comes close to the examples of strict MWF inequality of a 4-component genus 3 (alternating) link and a genus 6 knot found in [MP]. Evidence from some computations during its proof led us to conjecture an improvement of Ohyama's [Oh] braid index inequality for special alternating links (§7.5). We confirm this stronger inequality (apart from knots of genus up to 4) also for arborescently alternating links.

Our study of Murasugi–Przytycki's index also requires us to clarify and correct an inexactness in [MP], which affects the proof of their main upper braid index estimate (see §7.2). We revealed a slight discrepancy between their definition of graph index and the diagrammatic move they introduce to reduce the number of Seifert circles. We will be able to justify their estimates, but still there is some cost in that the correspondence between a diagram and (its Seifert) graph becomes (in general) lost during the recursive calculation of the index (using their old definition). This oversight seems to propagate to other papers and may cause a problem at some point. We will explain how to define, on the level of Seifert graphs, the "right" index w.r.t. their move.

In Chapter 8 we develop a method of constructing a Bennequin surface of links for a minimal genus canonical surface. The number of strings of the resulting braid is generally low, and can be calculated by modifying our corrected version of graph index. We apply this construction to alternating knots of genus up to 4 and show that one can span for these knots a Bennequin surface on a minimal string braid. Bennequin [Be] proved such a result for all 3-braid links and Hirasawa (unpublished) for the 2-bridge links. In contrast, knots lacking such surfaces are known for braid index 4 and genus 3. We will see how to apply the Murasugi–Przytycki move keeping a minimal genus surface, finally converting it into a minimal string Bennequin surface.

1.5.7 Properties of the Alexander polynomial of alternating knots

The final research Chapter 9 deals with the Alexander polynomial of an alternating knot. We consider first a conjecture of Hoste about the roots of the Alexander polynomial. We develop several tests based on the generator description, and apply them to confirm this conjecture on knots up to genus 4 (Theorem 9.1.2).

Another conjecture we treat is the log-concavity conjecture of [St16]. It states that the square of each coefficient of the Alexander polynomial of an alternating link is not less than the product of its two neighbors. This implies Fox's "trapezoidal" Conjecture 9.2.2. We will verify both conjectures for knots up to genus 4 (Theorem 9.2.5). In this case, it was possible to extend the result to links (Theorem 9.2.7).

As another qualitative extension of the Fox conjecture, we will conclude that the convex hull of alternating knot polynomials of a given genus is a polytope, and we can determine this polytope (i.e., the complete set of linear inequalities satisfied by the coefficients of Alexander polynomials of alternating knots) up to genus 4 in §9.3. Apart from an improvement of the trapezoidal inequalities, we will compare our result also to the inequalities obtained by Ozsváth–Szabó using knot Floer homology [OS]. (Their inequalities give an alternative proof of the trapezoidal conjecture for genus 2.) In-Dae Jong [Jn] used the generator description (Theorem 2.11.11) to prove the log-concavity conjecture and then, following the discussion in §9.3, to give the complete set of linear inequalities for genus 2.

1.6 Issues of presentation

As in a research monograph, material was and is prepared to be used extensively in other publications, and its correctness is critical. Thus, statements are generally provided with proofs, some of which are long and technical. Furthermore, for the same reason, I do not include any exercises. In that regard, I am haunted by the words of a referee very early on in my publishing career. I find them important not only for my own case, and thus I try to reproduce them here as exactly as I can recall:

> And there is one final remark. There are statements in this paper labeled "Exercise." According to a tradition established recently among [...] knot theorists, this tends to mean "a statement which I do not know how to prove, but which will be used in the following." If this is not the case, then the status of such an exercise should be explained: Is it a theorem, a question, or a conjecture? Sometimes even something like "it is easy to see..." is acceptable, because in that case the author takes responsibility for the correctness of the statement. But who is responsible for an exercise? Maybe the reader who was not able to solve it?

A substantial part of the work relies on computation. In these cases, input and output will be explained, as well as certain algorithmical details relevant for the mathematical background. However, there will be little computer science discussion of how algorithms were implemented in a concrete language (I used partly Mathematica [Wo] and partly C++), excerpt of code, etc. I can, though, of course, share the relevant code at request.

1.7 Further applications

The last chapter is again written in a somewhat more relaxed style as it is provided to place the content in this book within an even larger body of work related to "generator theory." This refers both to previous papers, as well as announcements how the research here will be used subsequently.

We will, among other things, discuss the relation to triangulations of surfaces and Wicks forms, fiberedness, contact geometry invariants, hyperbolic volume, and mention very briefly some further topics.

Chapter 2

Preliminaries

2.1 Knots and diagrams

A *knot* will be an embedding of the circle S^1 in S^3 (or $\mathbb{R}^3$), a *link* an embedding of a collection of circles. These will be called *components* of the link. Thus a knot is a link of 1 component. (Usually the case of knots is included in a statement about links, but it will sometimes depend a bit on the context of the statement so a little care may be necessary.) The *number of components* of a link L will be written as $n(L)$ (so $n(K) = 1$ if K is a knot). We will assume $n(L) \geq 1$, i.e., we will not treat the empty link. To avoid wild behavior, it suffices to assume, e.g., that knots and links are smooth and regarded up to smooth isotopy of the ambient space.

Knots and links are commonly represented by *diagrams*, which consist of smooth plane curves with *crossings*. Crossings are understood to be transverse intersections (self-intersections of a curve or intersections of different curves for links) with added over- and undercrossing strand information.

Let D be a diagram of a knot of link L. We write also $n(D) = n(L)$ for the *number of components* of D, and $c(D)$ for the *number of crossings* of D.

A crossing p in a link diagram D is called *reducible* (or nugatory) if it looks like the left side of Figure 2.1. We call D reducible if it has a reducible crossing; otherwise we call it *reduced*. The reducing of the reducible crossing p is the move depicted in Figure 2.1. Each diagram D can be (made) reduced by a finite number of these moves.

(2.1)

FIGURE 2.1: Eliminating a reducible crossing

In the following we assume that all diagrams are reduced unless otherwise stated.

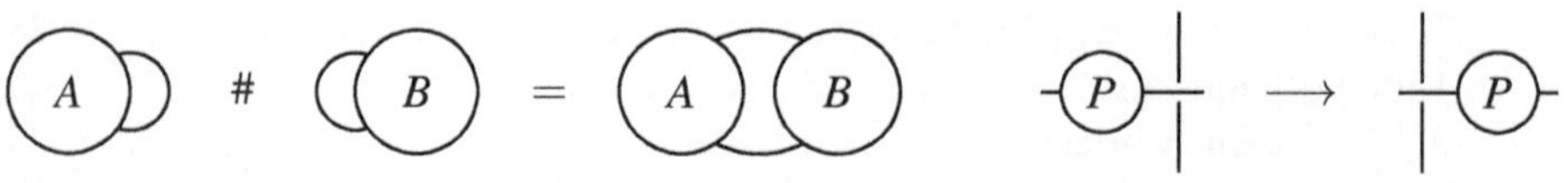

(*a*) (*b*)

FIGURE 2.2: Diagram connected sum and factor slide move.

Part (a) of Figure 2.2 displays the *connected sum* $D = A\#B$ of the diagrams A and B. The latter are called *factors* of D. The connected sum of diagrams is uniquely determined only up to the move shown in part (b) of the figure (and the mirror image of that move). We will call it a *factor slide*. If a diagram D can be represented as the connected sum of diagrams A and B, such that both A and B have at least one crossing, then D is called *composite*; otherwise it is called *prime*. A knot of link L is *prime* if whenever $D = A\#B$ is a composite diagram of L, one of A and B represent an unknotted arc (but not both; the unknot is not prime per convention).

A diagram D is *connected* if its curve is a connected set in $\mathbb{R}^2$, that is, there is no closed curve γ disjoint from D such that both the interior and exterior of γ have non-trivial intersection with D. Otherwise D is *disconnected* or *split*. A link is *split* if it has a split diagram and *non-split* if otherwise.

One of the classic primality criteria, for alternating diagrams, is due to Menasco.

Theorem 2.1.1 ([Me]) If D is a prime alternating non-trivial diagram of a link L, then L is prime.

2.2 Crossing number and writhe

The *crossing number* $c(K)$ of a knot K is the minimal crossing number $c(D)$ of all diagrams D of K.

This is a very basic quantity, but it is incredibly hard to determine in many situations and plagues many knot theorists. It is suggestive that crossing number is additive under connected sum,

$$c(K_1\#K_2) = c(K_1)+c(K_2)\,. \tag{2.2}$$

This is known for special classes (e.g., alternating knots), and a computer can confirm it for a myriad of instances. But the problem is open (and difficult) in general, and one among the many worthwhile and important problems too far afield from the focus of this text.

The *writhe*, or *(skein) sign*, is a number ± 1, assigned to any crossing in a link diagram. A crossing as in Figure 2.3(a) has writhe 1 and is called *positive*. A crossing as in Figure 2.3(b) has writhe -1 and is called *negative*. The *writhe* $w(D)$ of a link diagram D is the sum of writhes of all its crossings.

Let $c_\pm(D)$ be the number of positive, respectively negative crossings of a diagram D, so that $c(D) = c_+(D)+c_-(D)$ and $w(D) = c_+(D)-c_-(D)$. Let $c_\pm(K)$ for a knot K denote the minimal number of positive respectively negative crossings of a diagram of K.

FIGURE 2.3: A positive, negative, and smoothed out crossing.

A diagram is *positive* if all its crossings are positive. It is *almost positive* if all of its crossings are positive except for exactly one. A knot or link is *positive* if it has a positive diagram. See, e.g., [CM, O, Yo, Zu][1]. It is *almost positive* if it is not positive but has an almost positive diagram. More generally, a diagram D is *k-almost positive* if it has exactly k negative crossings, i.e., $c_-(D) = k$, and a knot K is k-almost positive if it has a k-almost positive but no $k-1$-almost positive diagram, i.e., $c_-(K) = k$. The switch of crossings of a diagram D so that all become positive will sometimes be called *positification* of D. Analogously one can define *negative* diagrams and knots.

A diagram is *almost alternating* [Ad, Ad3, Ad2, GHY] if it can be turned by one crossing change into an alternating one. A knot is *almost alternating* if it has an almost alternating diagram but is not alternating.

[1] Beware that some authors use the name for the narrower class of links with positive braid representations.

We write $!D$ for the *mirror image* of D. We obtain $!D$ by switching all crossings of D. Similarly $!K$ denotes the mirror image of K (i.e., $!K$ has the diagram $!D$, when K has the diagram D). Clearly $g(!D) = g(D)$ (and therefore $\tilde{g}(K) = \tilde{g}(!K)$) and $g(!K) = g(K)$.

Below we consider *amphicheiral*, or *achiral* knots. This is the designation of a knot which can be turned into its mirror image: $K = !K$. From the knots in Figure 1.1, the unknot is obviously amphicheiral. So is the figure-eight knot, as shows a simple exercise. In contrast, the trefoil is not amphicheiral. In other words, the left-hand trefoil and its mirror image, the right-hand trefoil on Figure 2.4, are two distinct knots. A knot like the trefoil which is not amphicheiral is called *chiral.*

2.3 Knotation and not-tables

Apart from their historic role, knot tables serve, also for us below, as an indispensable source for examples and counterexamples. In practice, it turns out useful to tabulate, for a given crossing number, only prime knots. (Note that this release from duty bears to some extent the ascertaining of (2.2).) Further, one usually disregards orientation of the knot and treats each knot only up to taking the mirror image (i.e., among a knot and its mirror image, only one is included).

Nowadays a prime knot is typically denoted as X_Y, where X is the crossing number, and Y is the order in the table. For example, there is only one (prime) knot of crossing number 3 (up to mirror image), the trefoil, and consequently it will be written 3_1. Similarly 4_1 is the figure-8-knot. But for 5 or more crossings, there is more than one prime knot, so the question is: What is the order? However commendable many tabulators' efforts, it seems also necessary to recount some of the pitiful history on this very basic issue.

According to the goal of a table, knots included therein more than once (duplications) or not at all (omissions) constitute errors. The compilation (notably in the early stages) and presentation of these tables has propagated not only such errors, of which the Perko pair [Pe] is likely only the most notorious one, but even errors in the errors [E]. The origin of this particular plague is some discord over the last four knots in Rolfsen's tables [Ro, appendix] after Perko's duplication emerged. Their indices are often shifted down by 1. Some record on this and other problems in the table (like the mix-up of 10_{83} and 10_{86}) can be found on Bar-Natan's Knot Atlas [BN].

Even if correct, different compilation procedures will, of course, render the list in a different order. Apparently, the attempts to popularize one's own (or, at least, the failure to coordinate each other's) ordering, and to disregard previous conventions, have a tradition almost as long as tabulation itself. Let me pay tribute to a senior who should know better and whose discussion [Pe3] is unpublished, but I find worth hearing.

Note that Reidemeister [Re], following Alexander and Briggs [AB], does not use Tait's original numbering [T]—their 9_{23} is Tait's 9-XII. So Bar-Natan was not the first Princeton-educated topologist to tamper[2] with the previously well established numbering of tabled knots (see [E] and [St22] for consequences). Others have also chosen to ignore the historic nomenclature handed down by Tait, Little, and Conway [Ro, HTW] and, alas, our own efforts to preserve the established order in papers classifying the 10- and 11-crossing knots [Pe, Pe4] have long been consigned to the dust bin of history, most recently under the rubric of "old Perko's notation" [CDM, p. 49].[3]

Rolfsen's tables of ≤ 10 *crossing knots* (which accord with Reidemeister and Alexander-Briggs) are now widely accepted at least because of the importance his book has gained. We thus feel compelled to stick to this quasi-standard here (and are lucky to mostly avoid its described problems).

For *knots from 11 to 16 crossings* our numbering is that of the tables *included in KnotScape* [HT]. This choice is simply because I largely appealed to KnotScape's functionality for diverse computations. However, I reorganize the KnotScape tables so that non-alternating knots are appended after alternating ones (of the same crossing number), instead of using "a" and "n" super/subscripts. For example, there are 367 prime alternating knots of 11 crossings. Thus a knot commonly named 11_{a111} in the table will be written as 11_{111}, while 11_{n111} as 11_{478} (for $111+367=478$).

Knot diagrams of 17 and 18 crossings were generated by the algorithm (and software) of [RFS], and are applied for some computational checks, but a numbering is never used.

As a noteworthy sequel to the above story of Perko, be aware that Hoste–Thistlethwaite break even with the Rolfsen numbering for ≤ 10 crossing knots, and they do so without clear warning. (Already the two 5 crossing knots are exchanged.) Visible bruises of the obstinacy protrude on [Kf4, p. 218] (where 10_{48} is written as 10_{79}) and in the knot tables of the very first edition of Adams' Knot book [Ad]. For my own use, I manually recoded a part of KnotScape; I saw this as the least awkward way of preventing more annoying miscommunication. See [An] for an entire book on tabulation yet completely independent from [HTW] and previous efforts.

As a further small twist, it will be necessary for us to distinguish at a few places below between a knot and its mirror image. We thus have to fix which knot should be X_Y and which $!X_Y$. We will use the *mirroring convention in KnotScape's tables throughout*. It is not easy to describe in a few words, but one distinctive feature is that all diagrams there have at least one positive crossing. This explains the choice for all

---

[2] To set the record straight: Bar-Natan merely clarifies and follows a practice (regarding the last four Rolfsen knots) initiated long before, notably in [BZ], and reiterated since by many others (incl. myself).

[3] Besides all his valid criticism, one should also take into account that Perko, as a hobby-mathematician (and professional lawyer), achieved a publishing record enviable to many of us so-called mathematics "professionals" (incl. myself).

knots occurring there with positive diagrams, which includes all special alternating knots, and should suffice here.

Our use of mirroring for ≤ 10 crossing knots means that a knot X_Y in Rolfsen's tables will be mirrored for us the way it occurs in KnotScape's tables. Most importantly, 3_1 will be thus the positive (right-hand) trefoil, while $!3_1$ the negative (left-hand) one—*unlike* in Rolfsen's table. Keep in mind that the knot usually appears in KnotScape at a different place. Thus, for instance, 5_1 will be for us the $(2,5)$-torus knot mirrored positively, even although it is mirrored negatively by Rolfsen, and written as 5_2 by Hoste–Thistlethwaite. A detailed conversion list can be found on my website [St23].

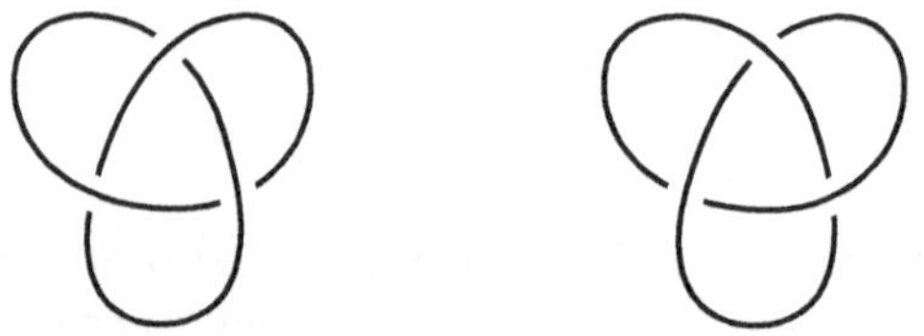

left-hand trefoil $!3_1$ right-hand trefoil 3_1

FIGURE 2.4: The two trefoils

Caution—and, desirably, explanation—of what one uses and how thus becomes a to-be-present—even if not pleasant—companion of taking advantage of knot tables. (The case of links is a Pandora's box I do not even dare to open...) I personally find history deserves some care to be maintained, even by champions, rather than the amount of confusion, hassle, and dismay that results from taking it less seriously. Perko finishes his note by quoting Reidemeister [Re2, p. 92]: "Das mathematische Denken is nur der Anfang des Denkens."[4]

2.4 Seifert surfaces and genera

A *Seifert surface* of an oriented knot or link L is a compact oriented surface S embedded in S^3 with boundary $\partial S = L$.

$$S \hookrightarrow S^3, \qquad \partial S = L. \tag{2.3}$$

Whenever S is oriented, there is an induced orientation on $\partial S = L$, and we will assume that this orientation coincides with the given orientation of L for each component of

[4]"Mathematical thinking is only the beginning of thinking."

L. It is possible that S is disconnected (although these cases will not be of much interest here; see Theorem 2.8.1 for the reason). However, we will assume that no component of S is closed, that is, each component of S bounds at least one component of L.

By the classification of compact oriented surfaces, a connected surface is determined by its *genus* $g = g(S)$ (pictorially, the number of holes of S) and the number of *boundary components* n.

The example for $n = 1$ and $g = 2$ is:

$S =$ 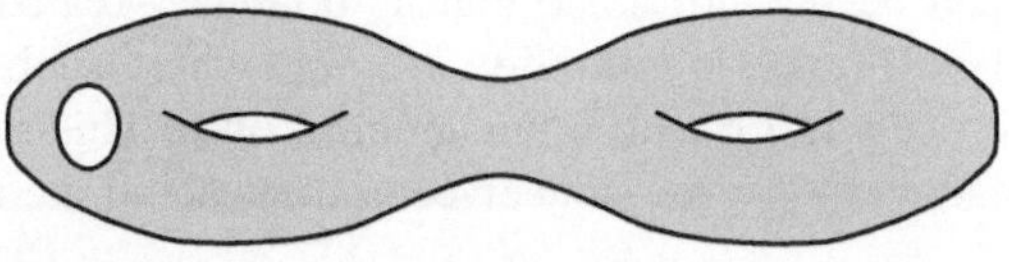 (2.4)

The *Euler characteristic* of S is

$$\chi(S) = 2 - 2g - n\,. \tag{2.5}$$

The *(Seifert) genus* $g(L)$ of a knot or link L is given by

$$g(L) = \min\{\, g(S) : S \text{ is a Seifert surface of } L\,\}\,.$$

The unknot is the only knot of genus 0. Similarly, the *Euler characteristic* $\chi(L)$ is said to be the maximal Euler characteristic of Seifert surface of L. It is immediately clear from (2.5) that when L bounds no disconnected Seifert surface, then

$$2g(L) = 2 - n(L) - \chi(L)\,.$$

For a diagram D of L, there is a procedure of Seifert [Se], called *Seifert algorithm*, to construct such a surface whenever a diagram D of L is given. The procedure is briefly this.

The replacement of a positive or negative crossing (a) respectively (b) in Figure 2.3 by the no-crossing fragment (c) appearing on the right will be called *smoothing out* the crossing. First, smooth out all crossings of D (but keep information on where crossings were located). The resulting disjoint plane loops are called the *Seifert circles* of D. Then, plug disks into all Seifert circles, and finally connect these disks by half-twisted bands to reinstall the crossings (at the place they were originally smoothed out).

After applying this procedure, we now observe the knot or link depicted in D to carry a surface, which is easily seen to be orientable (and respecting the orientation of L). Such a Seifert surface will be called *canonical* (for the diagram D) and will play a major role below.

The *(canonical) genus* $g(D)$ of D is defined to be the genus of the Seifert surface obtained by Seifert's algorithm on D, and its Euler characteristic $\chi(D)$ is called *(canonical) Euler characteristic* (of D).

It is obvious that for every diagram D of L,

$$g(D) \geq g(L), \quad \text{and} \quad \chi(D) \leq \chi(L).$$

If equality holds, we will say that D has *minimal genus*. It should be noted that L possesses a multitude of Seifert surfaces, and not all of them occur as canonical surfaces for some diagram of L. Morton [Mo] has found the first examples of knots where in fact no canonical Seifert surface has minimal genus.

It should be clear that the embedding (2.3) will hardly be as neat as in (2.4). The surface in (2.4) can be set into S^3 so that its boundary realizes any genus 2 knot. Theorem 2.11.11 will reveal a multitude of knots where such surfaces are canonical, and there are many more with non-canonical ones. This serve as a caveat to an unexperienced reader against pre-conceived notions about the appearance of Seifert surfaces.

Note also that a surface S as in (2.3) is obviously not of minimal genus because "a handle can be cut off." This can be made more precise by saying that there is a loop $C \subset S$ bounding a disk in the complement of S that is not isotopable onto S. Such surfaces are called *compressible* and the others *incompressible*. A minimal genus surface must obviously be incompressible, but beware that the converse is not always true.

Write $s(D)$ for the *number of Seifert circles* of D. Then, with $c(D)$ the number of crossings, and $n(D) = n(L)$ the number of components as in §2.1,2.2, we have

$$\chi(D) = s(D) - c(D) \text{ and } 2g(D) = 2 - n(D) - \chi(D).$$

Theorem 2.4.1 (see [Mu2, Cw, Ga, Cr]) If D has an alternating or positive diagram of L, then $g(L) = g(D)$ and $\chi(L) = \chi(D)$.

The *canonical genus* $\tilde{g}(L)$ respectively *canonical Euler characteristic* $\tilde{\chi}(L)$ is defined as the minimal genus respectively maximal Euler characteristic of all diagrams of L. As cautioned by Morton's examples (see [Mo]), in general we can have $g(L) < \tilde{g}(L)$.

Bennequin's inequality (theorem 3 in [Be]) can be stated, using (as explained in [St]) the work by Vogel [Vo], in the following way.

Theorem 2.4.2 If D is a diagram of a knot K, then $g(K) \geq g(D) - c_-(D)$.

In particular, if D is a diagram of the unknot, then $c_-(D) \geq g(D)$.

We call a crossing p connected to a Seifert circle s also *adjacent* or *attached* to s. The *valency* of a Seifert circle s is the number of crossings attached to s. We call a Seifert circle *negative* if only negative crossings are attached to it. Let $s_-(D)$ be the number of negative Seifert circles of D.

Bennequin's inequality was improved by Rudolph [Ru]. We state this improvement only for knots. Because a knot K lives in S^3, which bounds the 4-ball B^4, there are (smooth) surfaces bounding K in B^4. The *smooth slice genus* $g_s(K)$ of K is the minimal genus of such a surface. Every Seifert surface can be pushed into B^4, and thus

$$g(K) \geq g_s(K)$$

for every knot K. A knot K is *(smoothly) slice* if $g_s(K) = 0$.

Theorem 2.4.3 If D is a diagram of a knot K, then

$$g(K) \geq g_s(K) \geq g(D) - c_-(D) + s_-(D) . \tag{2.6}$$

We will refer to (2.6) as the *Rudolph–Bennequin inequality*.

The notion of Seifert surfaces is fundamental, and every (serious) general knot theory textbook should cover it in some way, as well as at least also mentioning Seifert's method. The topology of Seifert surfaces is discussed at length in [Ro]. For a more combinatorial point of view, which is more useful here, see [Cr2, Chapters 2 and 5]. A basic treatise can also be found in Adams' book [Ad].

2.4.1 Special diagrams

A diagram is *special* if no Seifert circle contains other Seifert circles in both regions that it separates the plane into. Such Seifert circles are called *separating*.

Let us here for the sake of thoroughness briefly outline the reason for the following well-known but useful fact.

Lemma 2.4.4 For a connected diagram D, two of the properties special, positive/negative, and alternating imply the third.

Proof. We will show that a positive connected diagram D is alternating if and only if it is special. This argument will make clear how to obtain the rest of the statement.

Thus let D be connected and positive.

First, assume D has a separating Seifert circle s, i.e., D is not special. Because D is connected, there are crossings attached to s from inside and from outside. Let p, q be two consecutive crossings in cyclic ordering along s, such that p is attached to s from inside and q from outside. Then the overpassing strand of p is also an overpassing strand of q, and thus D is not alternating.

Now, conversely assume D is special. Assume a strand passes a crossing p from below. Then it becomes part of a Seifert circle s to which p is attached. The crossing q next to p in cyclic ordering along s is attached to s from the same side as p (because D is special and s is not separating). Thus the underpassing strand of p exits s as an overpassing strand of q. Consequently, every underpass in D is followed by an overpass, and one easily sees also that every overpass is followed by an underpass. Thus D is alternating. □

A diagram with the three properties in Lemma 2.4.4 is called *special alternating*. A knot is special alternating if it has a special alternating diagram. Such knots were introduced and studied by Murasugi [Mu] and have a series of special features. Conversely all knots have a special (not necessarily alternating) diagram. Hirasawa [Hi2] shows how to a modify any knot diagram D into a special diagram D' so that $g(D) = g(D')$. (Actually, the canonical surfaces of D and D' are isotopic.) We will study his procedure in great detail below (§3.2).

The *index* $\operatorname{ind}(s)$ of a separating Seifert circle s is defined as follows: Denote for an inner crossing (i.e., attached from the inside) of s a letter i, and for an outer crossing a letter o cyclically along s. Then $\operatorname{ind}(s)$ is by definition the minimal number of disjoint subwords of the from i^n or o^n $(n > 0)$ of this cyclic word. (Put differently, it is the number of pairs p,q as in the first part of the proof of Lemma 2.4.4.) For example, the Seifert circle s in Figure 3.1 on page 49 has index 4.

2.5 Graphs

2.5.1 General definitions

There are some close connections between knot and link diagrams and graphs, which will also be very relevant here. One such is the Seifert graph (see §2.9). We must thus spend effort very early on to give some basic preliminaries on graphs.

Note: Seifert graphs are one important occurrence of graphs in knot theory but not the only one by far. Another is the checkerboard graph, which we will use (§7.5) and thus discuss (§2.10) more briefly. For other graph-theoretic excourses, see, e.g., [CDM]. In §10.6 we will sketch some possible connection between the occurrence of trivalent graphs in Vassiliev invariant theory and our context.

A graph G will have for us possibly multiple edges (edges connecting the same two vertices) but usually no loop edges (edges connecting one and the same vertex). Edges connecting the same two vertices will be called *parallel*.

By $V(G)$ we will denote the set of vertices of G, and by $E(G)$ the set of edges of G, with each multiple edge counting as a set of single parallel edges; $v(G)$ and $e(G)$ will be the number of vertices and edges of G (thus counted), respectively.

The number of edges with endpoint $v \in V(G)$ is called the *valence* $\operatorname{val}_G(v)$ of v. A vertex of valence 0 is called *isolated*.

For a graph, let the operation

(adding a vertex of valence 2) be called *bisecting* and its inverse (removing such a vertex) *unbisecting* (of an edge). If G has no vertices of valence 2, we call a graph G' a *bisection* of G if G' is obtained from G by a sequence of edge bisections. We call a bisection G' *reduced* if it has no adjacent vertices of valence 2. (For us usually G will be a trivalent graph, and a reduced bisection G' of G means that each edge of G is bisected at most once to obtain G'.) Conversely if G' is a graph, its *unbisected graph* G is the graph with no valence-2 vertices, of which G' is a bisection.

Similarly, a *contraction* is the operation

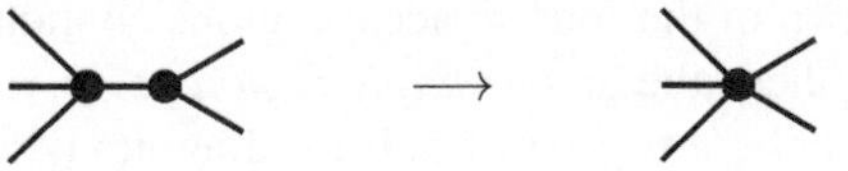

,

and a *decontraction* its inverse.

The *doubling* of an edge consists in adding a new (parallel) edge connecting the same two vertices.

A graph G is *series-parallel* if it can be obtained from  by edge iterated bisections and doublings.

A graph G is *connected* if for any two vertices $v, w \in V(E)$ there is a *path* connecting them, i.e., there are vertices $v_0 = v$, $v_1, \dots, v_n = w$ with an edge between v_i and v_{i+1} in G. A graph which is not connected is *disconnected*.

A *cycle* in a graph should be defined in our situation with a bit of caution to allow the following three options: it is (a) a single loop edge (will usually not occur), (b) a collection of two parallel edges, (c) a set of $n \geq 3$ edges e_i, $i = 1, \dots, n$ such that there are vertices $v_1, \dots, v_n$ of G with e_i connecting v_i and $v_{i \bmod n+1}$.

A graph is a *forest* if it has no cycles. It is a *tree* if it has no cycles and is connected. (According to this description, an isolated vertex is a tree.)

A graph is *n-connected* if n is the minimal number of edges needed to remove from it to disconnect it. (Thus connected means not 0-connected.) Such a collection of edges is called an *n-cut*.

From now when we delete an edge, we understand that a vertex it is incident to is *not* to be deleted too. In that sense, the set $\operatorname{star} v$ of edges incident to a given vertex v, called *star* of v, always forms a cut—if we delete these edges, v becomes an isolated component; thus the graph is not connected anymore. A *cut vertex* is a vertex which disconnects a graph when removed *together* with all its incident edges.

A graph G is *planar* if it is embeddable in the plane *and equipped* with a fixed such planar embedding. Observe that there is a natural bijection of edges between a planar graph G and its *dual graph* G^*; in that sense we can talk of the dual $e^* \in E(G^*)$ of an edge $e \in E(G)$. The operations doubling and bisecting become dual to each other.

A *subgraph* Γ of G is a graph Γ with $E(\Gamma) \subset E(G)$ and $V(\Gamma) = V(G)$. (Thus, according to this definition, we may have $\operatorname{val}_\Gamma(v) = 0$ while $\operatorname{val}_G(v) > 0$.) A graph Γ is a *spanning tree* of G if Γ is a subgraph of G and Γ is a tree.

A graph G is *bipartite* if there is a disjoint division $V(G) = V_1 \cup V_2$ (with $V_1 \cap V_2 = \varnothing$), such that each edge in G connects $v \in V_1$ with $w \in V_2$.

2.5.2 The checkerboard coloring and checkerboard graph

Definition 2.5.1 A *region* of a link diagram is a connected component of the complement of the (plane curve of) the diagram. An *edge* of D is the part of the plane curve of D between two crossings (clearly each edge bounds two regions). At each crossing p, exactly two of the four adjacent regions contain a part of the Seifert circles near p. We call these the *Seifert circle regions* of p. The other two regions are called the *non-Seifert circle regions* of p. If the diagram is special, each Seifert circle coincides with (the boundary of) some region. We call the regions accordingly Seifert circle regions or non-Seifert circle regions (without regard to a particular crossing).

Note that a connected diagram of n crossings has $n+2$ regions.

To any link diagram D one can associate its *checkerboard graph* $G = G(D)$ in the following manner. Consider the plane curve $\hat{D}$ of D as a 4-valent planar graph.

A *checkerboard coloring* of a diagram D is obtained as follows. We assign to regions (connected components of the complement) of the diagram D in the plane colors black and white so that regions bounded by a common edge of D have opposite color. Let $B(D)$ be the number of black and $W(D)$ of white regions. The total number of regions is

$$W(D) + B(D) = c(D) + 2. \tag{2.7}$$

Then $G(D)$ is defined to have a vertex for each black region of $\hat{D}$ and an edge for each crossing. See Figure 2.5 (or also, e.g., [Th4, Kf2]). We should note that $G(D)$ is in fact a *planar* graph, i.e. comes with a planar embedding.

To avoid pathologies occurring with $G(D)$, we will assume from now on that a link diagram D is connected and hence so is $G(D)$. This is automatically true for knot diagrams, but in the few situations where we discuss links, we will consequently assume that links are non-split.

When D is connected, then a switch between color in the checkerboard coloring of D results in taking the dual of the checkerboard graph. This graph is thus defined up to duality.

A diagram D is *arborescent* if its checkerboard graphs are series-parallel (see e.g., [St17]); it is easy to see that the property series-parallel is preserved under duality. A knot or link is arborescent (or Conway-algebraic[5]) if it has an arborescent diagram.

The class of arborescent links contains the *Montesinos links*, which in turn contain the *2-bridge (or rational)* links. These classes are a little more difficult to describe in the present setting and are better characterized in terms of their Conway notation

[5] Beware that there is another, and totally different, class of knots named "algebraic"!

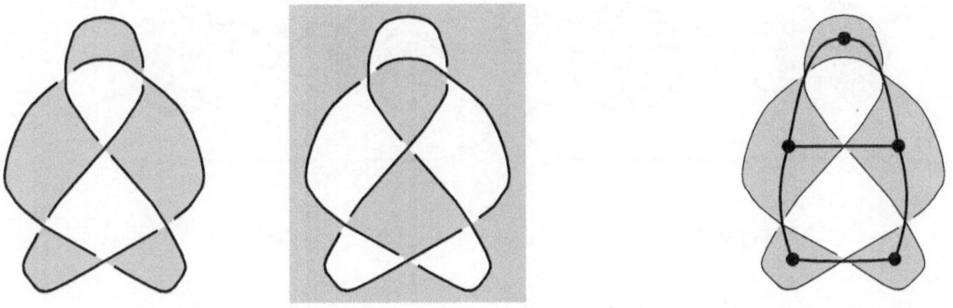

two checkerboard colorings

the checkerboard graph of the first coloring

FIGURE 2.5: Checkerboard coloring and checkerboard graph.

[Co]. A good account on Conway's notation (at least for arborescent links) is given in [Ad, §2.3]. For a more advanced treatment of Montesinos and 2-bridge links, see [BZ]. Montesinos and 2-bridge links will play a comparatively minor role below, and thus we prefer to leave the reader with these references instead of opening yet another thread here.

2.6 Diagrammatic moves

Moves between link diagrams, not necessarily preserving the underlying link, will play an important role. We already learned about those of Reidemeister in §1.2. Here we discuss further moves which will be needed later. (This list is not complete, one major addition being the $\bar{t}'_2$ twist introduced in §2.11.)

2.6.1 Flypes and mutations

Definition 2.6.1 A *flype* is a move on a diagram shown in Figure 1.2. We say that a crossing *admits* a flype if it can be represented as the distinguished crossing in the diagrams in the figure and both tangles have at least one crossing.

By the fundamental work of Menasco–Thistlethwaite, we have a proof of the Tait Flyping conjecture.

Theorem 2.6.2 ([MT]) For two alternating diagrams of the same prime alternating link, there is a sequence of flypes taking the one diagram into the other.

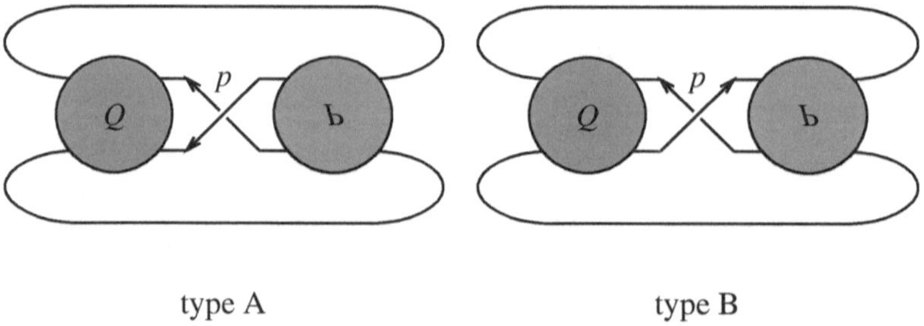

FIGURE 2.6: A flype of type A and B.

We introduced (see [SV]) a distinction of flypes according to the orientation near the crossing p at which the flype is performed. See Figure 2.6. An important observation is that each crossing admits at most one of the types A and B of flypes, and this remains so after applying any sequence of flypes on the diagram. We will elaborate more on this point in §2.11.

Mutation was introduced by Conway [Co]. It is a procedure of rotating by π a 2-string tangle in a tangle decomposition of a link diagram D along the axis perpendicular to the projection plane or horizontal or vertical in the projection plane. For example:

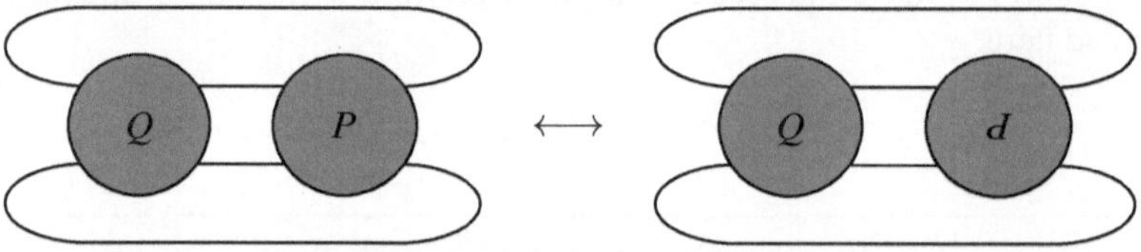

(2.8)

(In oriented diagrams D, one may have to change the direction of all strands of either P or Q to make in- and outputs compatible.) The sphere, intersecting the link in 4 points, which separates both tangles, is called *Conway sphere*.

Knots and links are *mutants* if they are transformable into each other by a sequence of mutations. Mutations are very similar to flypes, and in fact one easily sees that one can realize a flype by a sequence of mutations. However, in general mutations do change the link type. Nevertheless, mutants are "similar" and difficult to distinguish because most of the common (efficiently computable) invariants coincide on them, including the polynomial invariants we will introduce below. This failure to distinguish mutants, although disappointing at first, can be utilized to reduce a lot of work at several places (see §2.11). Mutations start for knots at 11 crossings, and the (historically) most famous example are the Kinoshita–Terasaka and Conway knots (see, e.g., [CM2, LM, MR]).

Menasco's work [Me] on Theorem 2.1.1 also shows that Conway spheres in alternating link complements are visible in an alternating diagram with a small family of exceptions, the so-called Menasco's *hidden Conway spheres*. One consequence worth mentioning is that if an alternating link has a mutant, that mutant is also alternating, and the mutation is visible in alternating diagrams. In particular, *the unknot has no mutants*, that is, a mutation applied on a diagram of the unknot will produce the unknot again (a circumstance known earlier). Lickorish's book [L, Chapter 4] contains more details.

2.6.2 Bridges and wave moves

A *bridge* is a piece of a strand of a knot diagram that passes only crossings from above. The *length* of the bridge is the number of crossings passed by it (*excluding* the initiating and terminating underpass). A *tunnel* is the mirror image of a bridge.[6]

A *wave move* is a replacement of a bridge a of length l_1 by another one b of length l_2. We will assume throughout that $l_2 \leq l_1$. The move is (crossing number) *reducing* if $l_2 < l_1$ (see Figure 2.7 or [StK] for example) and (crossing number) *preserving* if $l_2 = l_1$. Elsewhere a wave move is also called a (l_1, l_2)-*pass*. Note that a $(1,0)$-pass is the removal of a nugatory crossing (much like in Figure 2.1, except that P is flipped), and a $(1,1)$-pass is exactly the factor slide of Figure 2.2 (b). We will usually consider only reducing wave moves, except for the case $l_1 = l_2 = 1$. Obviously, such a move works for tunnels instead of bridges and we *will not distinguish between both* unless clearly stated.

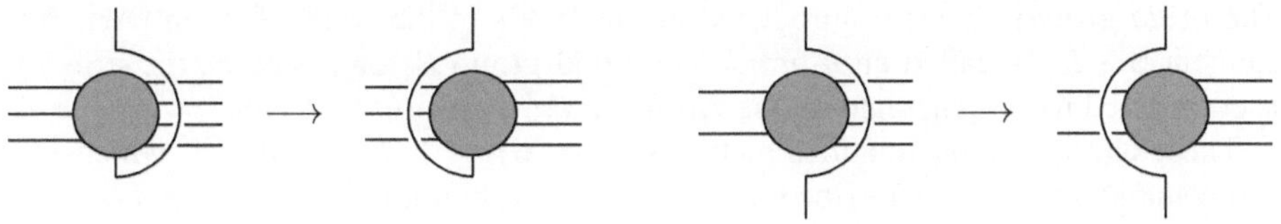

FIGURE 2.7: Wave moves. The number of strands on left and right of the shaded circle may vary. It is only important that the parities are equal respectively different, and that the left-outgoing strands are fewer that the right-outgoing ones.

Adams introduced a *tongue move* allowing us to create more complicated almost alternating diagrams from a given one. This move is shown in Figure 2.8. Herein the crossing needed to be switched to obtain an alternating diagram is encircled; we call this crossing *dealternator*.

Adams formulated a conjecture, stating how to recognize the unknot in almost alternating diagrams, which was proved recently by Tsukamoto.

Theorem 2.6.3 (Tsukamoto [Ts]) An almost alternating unknot diagram is trivializable by tongue moves, flypes, and crossing number reducing Reidemeister I and II moves.

[6] We confess to going here against established knot theory terminology with a different meaning. Exculpatorily, these classical sort of tunnels will never be treated here.

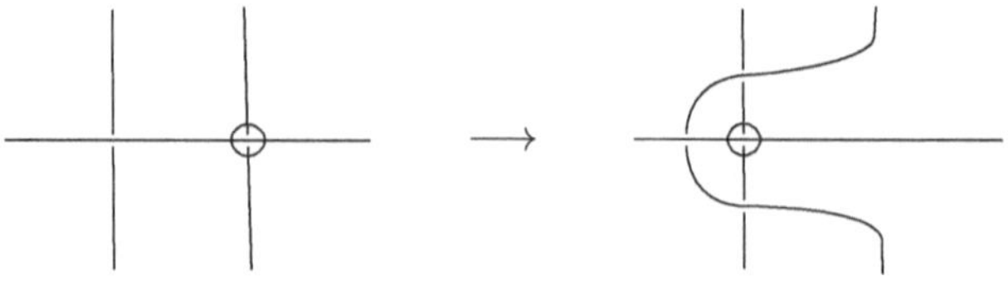

FIGURE 2.8: A tongue move between two almost alternating diagrams. The encircled crossing is to be switched to obtain an alternating diagram.

A tongue move is a special type of wave move, preserving the property almost alternating. In that sense Theorems 1.5.1 and 5.5.4 can be thought of as extending Tsukamoto's result in special diagrams.

2.7 Braids and braid representations

For braid-related knot theory, the basic reference remains Birman's book [Bi2].

The *braid groups* B_n were introduced in the 1930s in the work of Artin [Ar]. An element $\beta \in B_n$ is called an *n-braid*. The braid group B_n on n strands (or strings) is considered to be generated by the Artin *standard generators* σ_i for $i = 1, \ldots, n-1$. These are subject to relations of the type $[\sigma_i, \sigma_j] = 1$ for $|i-j| > 1$, which we call *commutativity relations* (the bracket denotes the commutator) and $\sigma_{i+1}\sigma_i\sigma_{i+1} = \sigma_i\sigma_{i+1}\sigma_i$, which we call *Yang-Baxter* (or shortly YB) relations.

There is a graphical representation of braids where in σ_i strands i and $i+1$ cross, and multiplication is given by stacking.

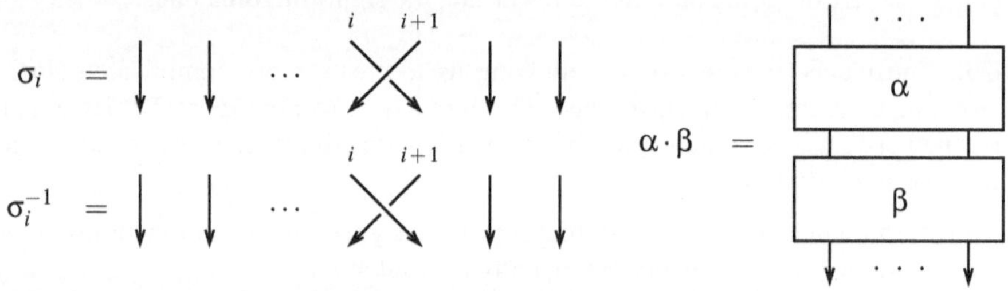

(Here we number strands from left to right and orient them downward. This convention, however, will not matter much.)

Alexander [Al2] related braids to links in real 3-dimensional space by means of a *closure* operation ˆ. In that realm, in became important to understand the *braid representations* of a given link L.

The *closure* $\hat{\beta}$ of a braid β is a knot or link (with orientation):

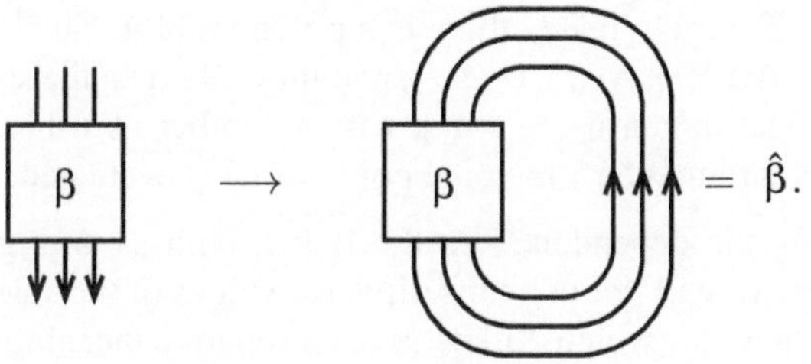

The *braid index* $b(L)$ of a link L is the smallest number of strands of a braid β whose closure $\hat{\beta}$ is L. See [Mo, FW, Mu4]. Alexander's theorem asserts that $L = \hat{\beta}$ for some braid β.

2.8 Link polynomials

2.8.1 Generalities

The purpose of this section is to give an introduction to link polynomials and their basic properties.

First, let us clarify the meaning of a "link polynomial." It is understood as an invariant of links

$$\{\text{knots and links}\} \to \text{some ring of polynomials}.$$

This means that every link now has its own polynomial. The ring of polynomials will have one or two variables, but we will assume that variables can occur in negative powers, i.e., consider *Laurent* polynomials. We will also allow half-integral powers for links. Coefficients will usually be integers.

The names of the variables will be specified for each invariant. Regarding the polynomials P and F we will introduce shortly, we mention already in advance that there are not only several variable names, but also *several other variable conventions*, differing from each other by possible inversion and/or multiplication of a variable by some fourth root of unity. We include some explanation how our conventions are related to those of some (mostly standard) literature, but the reader should be very careful about this point with comparisons to external sources.

Often variables of polynomials will be set to particular (most generally, complex, but most commonly, integer) values. If one fixes the variable so and varies the link, one then has a scalar (complex respectively integer) valued invariant.

In general we will take the liberty (or sloppiness) to alter notation of how the polynomial depends on the link and the variable. The one or the other will be omitted if they are suggestive from the context. For example, the Alexander polynomial Δ has the variable t. Thus for each link L, there is a polynomial in $\mathbb{Z}[t^{\pm 1/2}]$, which can be written as $\Delta(L)$, Δ_L, $\Delta(L)(t)$, $\Delta_L(t)$, or $\Delta(t)$, and they all mean the same thing. (There is obviously no danger in confusing a link with a number.) Further, one can replace a link L by a link diagram D on which the polynomial is evaluated.

Thus, we may write the dependence on the link L both as an argument, and as a subscript. However, we will *not* use subscripts for values of the variable: Thus, $\Delta(1)$ will never be written as Δ_1, which will be given a separate meaning later. Let us also agree that when the symbol i is used in an argument, *but not as a subscript*, it always denotes the complex unit $\sqrt{-1}$.

Before we introduce several link polynomials, we need some more notation regarding general polynomials. Let $X \in \mathbb{Z}[t,t^{-1}]$. The *minimal* or *maximal degree* $\min\deg X$ or $\max\deg X$ is the minimal respectively maximal exponent of t with non-zero coefficient in X. Let $\operatorname{span}_t X = \max\deg_t X - \min\deg_t X$, the *span* of *breadth* of X. The coefficient in degree d of t in X is denoted $[X]_{t^d}$ or $[X]_d$. The *leading coefficient* $\max\mathrm{cf}\, X$ of X is its coefficient in degree $\max\deg X$. If $X \in \mathbb{Z}[x_1^{\pm 1}, x_2^{\pm 1}]$, then $\max\deg_{x_1} X$ denotes the maximal degree in x_1. Minimal degree and coefficients are defined similarly, and of course $[X]_{x_1^k}$ is regarded as a polynomial in $x_2^{\pm 1}$.

2.8.2 Skein, Jones, and Alexander polynomial

The *skein polynomial* P [F&, LM, PTr] is a Laurent polynomial in two variables l and m of oriented knots and links and can be defined by being 1 on the unknot and the *skein* relation

$$l^{-1}P\big(\times\big)+lP\big(\times\big)=-mP\big(\,)\,(\,\big). \tag{2.9}$$

This equality relates the polynomials of the three fragments in Figure 2.3. We reiterate our understanding from §1.2 that these fragments represent triples of diagrams which are equal except at the depicted spot. We will often denote the diagrams in each triple as in (2.9) (from left to right) by D_+, D_- and D_0.

For a diagram D of a link L, we will use all of the notations $P(D) = P_D = P_D(l,m) = P(L)$ etc. for its skein polynomial, with the self-suggestive meaning of indices and arguments. The convention in (2.9) is similar to (and uses the same variables as) the one in [LM] but differs by the interchange of l and l^{-1}.

The *Jones polynomial* [J] V, and (one variable) *Alexander polynomial* [Al] Δ are obtained from P by the substitutions

$$V(t) = P(-it, i(t^{-1/2} - t^{1/2})), \tag{2.10}$$
$$\Delta(t) = P(i, i(t^{1/2} - t^{-1/2})), \tag{2.11}$$

hence these polynomials also satisfy corresponding skein relations, obtained by the same substitutions in (2.9).

The skein polynomial has at least 8 discoverers (in 5 teams): Freyd–Yetter, Hoste, Lickorish–Millett, Ocneanu, and Przytycki–Traczyk. These discoveries were almost simultaneous, but Przytycki–Traczyk's became known later due to the Marshall Law in Poland. There were attempts from very early on to name the polynomial by some acronym of the discoverers' names (with P and T included or not depending on the origin of the acronym). HOMFLY seems to have imposed itself over LYMFHO, FLYPMOTH, LYMPHTOFU (the latter, according to Bar-Natan [BN3], coined by Rudolph, with U for the Unknown further discoverers), and other imaginable pronounceable permutations of these letters. Acknowledging Przytycki–Traczyk's contribution, it is fairer to write HOMFLY-PT, the possibly best (and most widely used) acronym. There are also other names, like generalized Jones polynomial, Jones–Conway polynomial, 2-variable Jones polynomial, etc. Our choice of "skein polynomial" is pragmatic: it is impartial (and thus fair to all discoverers) and short (typing the 5 small letters of "skein" is a huge relief for one's fingers, as compared to 6 or 8 capital letters, in particular over the length of an entire book). Beware, though, that other authors carelessly assign the name "skein polynomial" also to the other invariants V and Δ which obey skein relations.

The advantage of a skein relation (2.9) is that, when systematically applied, it can recur the value of the skein polynomial of a link diagram to this of simpler diagrams. This gives a method for recursively calculating the polynomial for every link from a diagram and hence also for the Jones and Alexander polynomial.

Instead of Δ, we will sometimes also use the *Conway polynomial* [Co] $\nabla(z)$ with

$$\Delta(t) = \nabla(t^{1/2} - t^{-1/2}) . \tag{2.12}$$

Then ∇ satisfies the skein relation

$$\nabla(D_+) - \nabla(D_-) = z\nabla(D_0) .$$

It can be easily concluded from this relation that $\nabla(L)(z)$ is a genuine (and not Laurent) polynomial in z. Furthermore, it can be seen that $\nabla(L)(z) \in z^{n(L)-1}\mathbb{Z}[z^2]$; in particular, the Conway polynomial of a knot K satisfies

$$\nabla(K) \in \mathbb{Z}[z^2] . \tag{2.13}$$

Moreover, it is known that for all knots K,

$$[\nabla(K)]_0 = \nabla(0) = \Delta(1) = 1 . \tag{2.14}$$

Note that $\max\deg\nabla = 2\max\deg\Delta$, which we use in particular to implicitly restate some of the results of [Cr] in the sequel.

We will need several properties of the Alexander polynomial; here are two classical ones.

Theorem 2.8.1 If L is split, or more generally, bounds a disconnected Seifert surface, then $\Delta_L = 0$. If $\Delta_L \neq 0$ (and in particular L is non-split), then

$$2\max\deg\Delta_L \leq 1 - \chi(L) . \tag{2.15}$$

These are properties of the Alexander polynomial that are more obscured than elucidated by skein calculations. Theorem 2.8.1 originates from a connection between the Alexander polynomial and a Seifert surface (which is completely lacking for the skein and Jones polynomial). See, e.g., [Ro]. Kauffman's book [Kf3] treats (2.15) more extensively. We remark that, in Seifert-matrix-related algebraic topology, the Alexander polynomial usually emerges only up to units in $\mathbb{Z}[t,t^{-1}]$; the normalization coming from (2.11) is so that, for *knots* K,

$$\Delta_K(t) = \Delta_K(1/t) \text{ and } \Delta_K(1) = 1. \tag{2.16}$$

For links L, we have $\Delta_L(t) = (-1)^{n(L)-1}\Delta_L(1/t)$ and $\Delta_L(1) = 0$. There are also multivariable Alexander polynomials (for links) and higher order Alexander polynomials, but these will not be treated here.

Let us note in passing that for braid representations of links, there is yet another framework for calculating the Alexander polynomial from the Burau representation (and more generally the skein polynomial from Hecke algebra representations [J2]). Thus there are at least three different approaches to obtain and study properties of Δ:

- the skein relation,
- the Seifert matrix,
- the Burau matrix.

These connections are truly profound, and it is merely for the focus of the present book that we do not cover the second and third ones here, with [Ro] respectively [Bi2] already designated above as classical sources. A good further treatment of skein theory, from an originator himself, is given in [L].

Returning to (2.15), we notice that Crowell–Murasugi show the following further property for Theorem 2.4.1.

Theorem 2.8.2 (Crowell–Murasugi) If D is a non-split alternating diagram, then

$$2\max\deg\Delta_D = 1 - \chi(D). \tag{2.17}$$

The combination of these two theorems implies Theorem 2.4.1. It also implies:

Corollary 2.8.3 An alternating link L is non-split if and only if an alternating diagram D of L is non-split.

Moreover, it follows that (2.15) is an equality for non-split alternating links. Further, alternating links can be replaced in this discussion by positive links, and (2.17) for positive links is treated in [Cr].

For alternating links, Crowell–Murasugi proved one more property.

Theorem 2.8.4 ([Cw, Mu2]) The coefficients of Δ_K for an alternating knot K alternate in sign, i.e., $[\Delta_K]_i[\Delta_K]_{i+1} \leq 0$ for all $i \in \mathbb{Z}$.

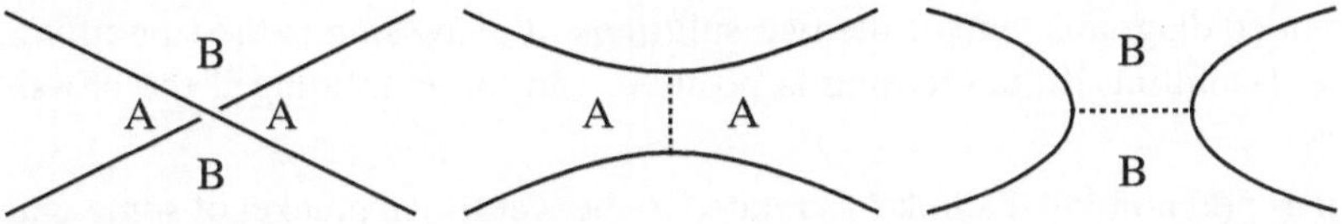

FIGURE 2.9: The A- and B-corners of a crossing, and both of its splittings. The corner A (respectively B) is the one passed by the overcrossing strand when rotated counterclockwise (respectively clockwise) toward the undercrossing strand. A type A (respectively B) splitting is obtained by connecting the A (respectively B) corners of the crossing. It is useful to put a "trace" of each split crossing as an arc connecting the loops at the splitted spot.

2.8.3 Kauffman polynomial and semiadequacy

The *Kauffman polynomial* [Kf] F is usually defined via a regular isotopy invariant $\Lambda(a,z)$ of unoriented links. We use here a slightly different convention for the variables in F, differing from [Kf, Th] by the interchange of a and a^{-1}. Thus in particular, we have for a link diagram D the relation $F(D)(a,z) = a^{w(D)}\Lambda(D)(a,z)$, where $\Lambda(D)$ is the writhe-unnormalized version of the polynomial, given in our convention by the properties

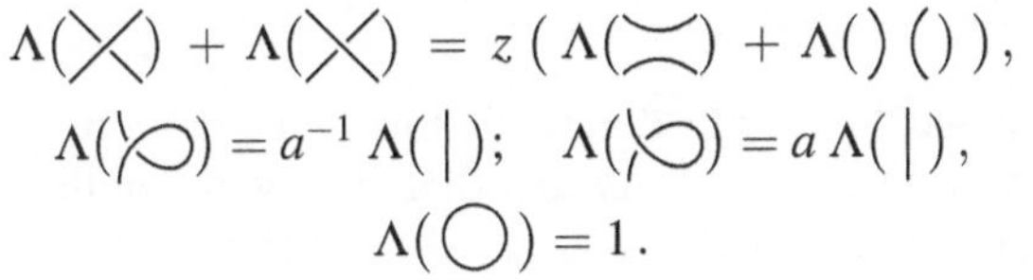

The Jones polynomial V is obtained from F (in our convention) by the substitution (see [Kf2, §III])

$$V(t) \quad = \quad F(-t^{3/4}, t^{1/4} + t^{-1/4}).$$

The *BLMH polynomial* [BLM, Ho] Q is defined by

$$Q(z) = F(1,z).$$

An alternative description of V is given by the Kauffman bracket in [Kf2], which we recall next. A *state* S is a choice of *splittings* of type A or B for any single crossing (see Figure 2.9). Let $\#A(S)$ and $\#B(S)$ denote the number of type A (respectively type B) splittings and $|S|$ the number of (disjoint) circles obtained after all splittings in S. The Kauffman bracket $[D]$ of a diagram D is a Laurent polynomial in a variable A, obtained by summing over all states S of D:

$$[D] = \sum_S A^{\#A(S)-\#B(S)} \left(-A^2 - A^{-2}\right)^{|S|-1}. \qquad (2.18)$$

Note that the description of splittings does not require orientation of the diagram.

For oriented diagrams, one of the two splittings of a crossing is the smoothing out; it is the A-splitting if the crossing is positive, and the B-splitting if the crossing is negative.

The Jones polynomial of a link L is related to the Kauffman bracket of some diagram D of L by

$$V_L(t) = \left(-t^{-3/4}\right)^{-w(D)} [D]\Big|_{A=t^{-1/4}} . \tag{2.19}$$

The first major application of the Jones polynomial (proved by Kauffman using the bracket) is the solution of Tait's Crossing conjecture.

Theorem 2.8.5 (Kauffman-Murasugi-Thistlethwaite) If D is a reduced alternating diagram of a link L, then $c(D) = c(L)$.

It will be useful to recall the following (see [St6]).

Theorem 2.8.6 An alternating diagram D represents a positive link L if and only if D is positive (and hence special alternating; recall §2.1).

Corollary 2.8.7 A link is special alternating if and only if it is positive and alternating.

Note that if we replace "link" by "diagram", this is the definition. However, a link may have diagrams with one property and diagrams with another, but no diagram that enjoys both properties simultaneously. The content of the above statement is that this phenomenon does not occur for the properties of (a diagram) being positive and being alternating.

Let the *A-state* $A(D)$ of D be the state where all crossings are A-splitted; similarly, define the B-state.[7] We call a diagram *A-(semi)adequate* if in the A-state no crossing trace (one of the dotted lines in Figure 2.9) connects a loop with itself. We call such a trace a *self-trace*. Similarly, we define B-(semi)adequate. A diagram is *semiadequate* if it is A- or B-semiadequate, and *adequate* if it is simultaneously A- and B-semiadequate. A link is adequate/semiadequate if it has an adequate/semiadequate diagram. Let us remark here that, while adequacy is only a slight extension of alternation, semiadequacy is a rather wide extension of adequacy. (Semiadequate knots contain completely positive, Montesinos and 3-braid knots.) Semiadequacy is still a fairly general condition, yet it helps settle many technical issues.

For semiadequate knots, the first coefficient of the Jones polynomial is ± 1, almost by definition [LT]. In an attempt to gain more insight into the appearance of the Jones polynomial, in [St5], and [DL, DL2], a detailed study was initiated of some coefficients of V in semiadequate diagrams, gaining an understanding of coefficients 2 and 3. This allowed us to prove, regarding Question 1.3.1:

[7] Note, however, that "$B(D)$" was allocated differently in §2.5.2; we will retain this meaning and never need a symbol for the B-state in the sequel.

Theorem 2.8.8 ([St11]) For a semiadequate knot (or link) diagram D, we have $V(D) \neq 1$ (or $V(D) \neq (-t^{1/2} - t^{-1/2})^{n(D)-1}$). That is, semiadequate knots and links have non-trivial Jones polynomial.

After the conversion (2.19), we see that the minimal degree of t to which the sum (2.18) contributes occurs in the A-state $A(D)$. For our subsequent arguments, this degree can be more conveniently written for a knot diagram D as

$$m(D) := g(D) - \frac{3}{2}c_-(D) + \frac{s(D) - |A(D)|}{2}. \tag{2.20}$$

If a diagram is A-adequate, only the A-state contributes in degree $m(D)$, and the coefficient is ± 1 (see [LT]). In Bae-Morton [BMo], the contribution of the sum of (2.18) in degree $m(D)$ for general diagrams D was studied. The following simple property will be useful.

Call a self-trace *isolated* if it does not pair up with another self-trace like

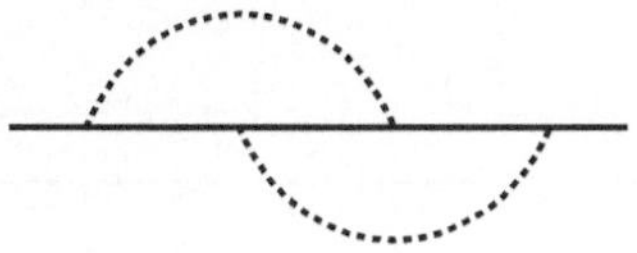

Lemma 2.8.9 (Bae-Morton [BMo]) If the A-state of D has an isolated self-trace, then the contribution of the sum of (2.18) in degree $m(D)$ vanishes; that is, $\min\deg V(D) > m(D)$.

Let $F(K) \in \mathbb{Z}[a^{\pm 1}, z]$ be the Kauffman polynomial of a knot K and $Q_K(z) = F_K(1,z)$ be the BLMH polynomial [BLM, Ho]. Let further

$$\begin{aligned} a_-(K) &= \tfrac{1}{2}\max\{m - l : [F(K)]_{a^l z^m} \neq 0\} \quad \text{and} \\ a_+(K) &= \tfrac{1}{2}\max\{m + l : [F(K)]_{a^l z^m} \neq 0\}. \end{aligned} \tag{2.21}$$

It easily follows that $\max\deg_z F(K) \leq a_+(K) + a_-(K)$. Moreover, if the r.h.s. is positive, strict inequality holds because the substitution $F(i,z) = 1$ shows that $\max\mathrm{cf}_z F(K)$ cannot be a single monomial (in a).

In [Th], Thistlethwaite proves that

$$c_\pm(K) \geq a_\pm(K), \tag{2.22}$$

with equality if and only if K is A, respectively, B-semiadequate. Let us say in the following that a (non-strict) inequality of the form '$A \geq B$' is *sharp* or *exact* if $A = B$ and *strict* otherwise (i.e., if $A > B$).

Adequate (in particular alternating) diagrams make both inequalities in (2.22) simultaneously sharp. It follows from these inequalities that for every non-trivial knot K,

$$\max\deg Q(K) \leq \max\deg_z F(K) < c_+(K) + c_-(K). \tag{2.23}$$

(If $c_-(K) = c_+(K) = 0$, then K is simultaneously positive and negative, and it is therefore trivial; see for example [St6], or the below remarks on the signature.)

It follows further that adequate diagrams are of minimal crossing number (thus extending Theorem 2.8.5).

One can also see that adequate diagrams of the same link have the same writhe. Note that for alternating diagrams, this writhe invariance follows from Theorem 2.6.2 because a flype preserves the writhe. It should be emphasized, though, that Theorem 2.6.2 is a very sophisticated tool, and the proof of writhe invariance using (merely) polynomial invariants must be considered to be simpler.

The writhe invariance has a somewhat illustrious history. I am grateful to K. Perko for its clarification. It appears, stated for arbitrary minimal crossing diagrams, as a *theorem* in Little's 1900 paper [Li] (communicated by Tait himself!), and that statement was later cemented by Dehn and Heegaard in their encyclopedia article [DH, p. 212]. This was thus apparently believed to be right for decades until it was surprisingly refuted by the Perko pair [Pe].

2.9 MWF inequality, Seifert graph and graph index

In [Mo, FW] it was proved that

$$\mathrm{mwf}(L) := \frac{1}{2}\,\mathrm{span}_l P(L) + 1 \le b(L)\,, \tag{2.24}$$

the *Morton–Williams–Franks* (MWF) *inequality*. Because we will need the left-hand side of this inequality later, let us write it as mwf(L) and call it the *Morton–Williams–Franks bound* for $b(L)$. The inequality (2.24) results from two other inequalities, due to Morton, namely that for a diagram D, we have

$$1 - s(D) + w(D) \le \min\deg_l P(D) \le \max\deg_l P(D) \le s(D) - 1 + w(D)\,. \tag{2.25}$$

Williams–Franks showed these inequalities for the case of braid representations. Later it was observed from the algorithms of Yamada [Ya] and Vogel [Vo] that the braid version is actually equivalent to—and not just a special case of—the diagram version. (These algorithms allow us to transform any diagram D into a braid diagram without altering $s(D)$ and $w(D)$.) Nonetheless we will refer below to (2.25) as "Morton's inequalities."

These inequalities were later improved in [MP] in a way that helps settling the braid index problem for many links (see Chapter 7 or also [Oh]). For this purpose, Murasugi–Przytycki developed the concept of *index of a graph*. We recall some main points of Murasugi–Przytycki's work, referring to [MP] for further details, and offer caution to the correction explained in [St20] and reviewed here in §7.1–7.3.

Definition 2.9.1 Let G be a signed graph (each edge carries a sign $+$ or $-$). For a vertex v in G let $G_v = G/v$ be the graph obtained by contracting the star of v, that is, the set of edges of which v is one of the endpoints (§2.5). Let $G \setminus v$ be the graph obtained from G by deleting all of these edges and v.

We say that v is a *cut vertex* if $G \setminus v$ is disconnected. A cut vertex decomposes G into a *block sum* or *join* $G_1 * G_2$ of two graphs $G_{1,2}$. If $G = G_1 * \cdots * G_n$ and all G_i have no cut vertex, the G_i will be called *block components* or *join factors* of G.

Definition 2.9.2 We define (recursively) a sequence of edges $\mu = (e_1, \ldots, e_n)$ to be *independent* in a graph G if the following conditions are satisfied.

1. The empty (edge) sequence is independent per definition.
2. Let e_1 connect vertices $v_{1,2}$. Then we demand that e_1 is *simple*, i.e., there is no other edge connecting $v_{1,2}$, and that $e_2, \ldots, e_n$ is independent in (one of) G_{v_1} or G_{v_2} (i.p., $e_2, \ldots, e_n$ are disjoint from the set of edges incident to v_1 or v_2 respectively).

An *independent set* is a set of edges admitting an ordering as an independent sequence.

The *index* $\mathrm{ind}\,(G)$, respectively *positive index* $\mathrm{ind}_+(G)$ and *negative index* $\mathrm{ind}_-(G)$ of G are defined as the maximal size/length of an independent edge set (or sequence), respectively independent positive or negative edge set/sequence in G. An independent set/sequence is *maximal independent* if it realizes the index of G.

(This index is to be kept strictly apart from the index of a Seifert circle of §2.1. Both are not unrelated, but we do not investigate their relation and use them in quite separate contexts.)

Now to each link diagram D we associate its *Seifert graph* $G = \Gamma(D)$, which is a planar bipartite signed graph. It consists of a vertex for each Seifert circle in D and an edge for each crossing, connecting two Seifert circles. Each edge is signed by the skein sign of the crossing it represents. We will for convenience sometimes identify crossings/Seifert circles of D with edges/vertices of G. Let also $\mathrm{ind}_{(\pm)}(D) = \mathrm{ind}_{(\pm)}(\Gamma(D))$.

Proposition 2.9.3 (see [MP, (8.4) and (8.8)]) If D is a diagram of an oriented link L, then

$$\begin{aligned} \max\deg_l P(L) &\leq w(D) + s(D) - 1 - 2\,\mathrm{ind}_+(D) && (2.26)\\ \min\deg_l P(L) &\geq w(D) - s(D) + 1 + 2\,\mathrm{ind}_-(D) && (2.27)\\ b(L) &\leq \mathrm{mpb}\,(D) := s(D) - \mathrm{ind}\,(D)\,. && (2.28) \end{aligned}$$

For any diagram D, we have

$$\mathrm{ind}_+(D) + \mathrm{ind}_-(D) \geq \mathrm{ind}\,(D)\,. \qquad (2.29)$$

For alternating (and more generally homogeneous [Cr]) diagrams D, equality holds because each join factor of $\Gamma(D)$ contains only edges of the same sign. This implies that if for such diagrams (2.26), (2.27) are sharp, then (2.24) and (2.28) become sharp, too.

It is possible to partially reconstruct D from $\Gamma(D)$ and completely when D is special and $\Gamma(D)$ is given in a planar embedding. If $\Gamma(D)$ has block components (Definition 2.9.1) G_i, then we call the (special) diagrams D_i with $\Gamma(D_i) = G_i$ the *Murasugi atoms* of D. Note that each separating Seifert circle of D gives a cut vertex of $\Gamma(D)$. The join operation of G_i at these cut vertices corresponds to an operation of identifying two Seifert circles in D_i, which is called planar (or diagrammatic) *Murasugi sum* (or $*$-product; see, e.g., [MP2]). Thus an alternative way to specify the Murasugi atoms D_i is to say that they are the prime factors of the Murasugi summands of D (see also Definition 3.2.1 below, and [QW]). However, the information of how under a Murasugi sum in D crossings attached to a Seifert circle on either side are arranged relatively to each other is lost in $\Gamma(D)$.

The following theorem of Murasugi–Przytycki is important:

Theorem 2.9.4 (Theorem 2.5 in [MP]) The index is additive under block sum of bipartite graphs, i.e., $\mathrm{ind}\,(G_1 * G_2) = \mathrm{ind}\,(G_1) + \mathrm{ind}\,(G_2)$ if $G_{1,2}$ are bipartite.

We will treat below, simultaneously with (the sharpness of) MWF, the following conjecture of theirs.

Conjecture 2.9.5 (Murasugi–Przytycki) If D is an alternating diagram of a link L, then $b(L) = \mathrm{mpb}\,(D)$.

Remark 2.9.6 We observed that for a special diagram D, Seifert circles coincide with the boundary of a collection of regions. It is easy to see that these regions do not share an edge, and thus they are the regions receiving the same color in a checkerboard coloring of D. This establishes that *for a special diagram D, the Seifert graph $\Gamma(D)$ coincides with one of the checkerboard graphs of D* (see §2.5.2).

2.10 The signature

The *signature* σ is a $\mathbb{Z}$-valued invariant of knots and links. Originally it was defined in terms of Seifert matrices [Ro]. We have that $\sigma(L)$ has the opposite parity to the number of components of a link L, whenever the *determinant*

$$\det(L) := \left|\Delta_L(-1)\right| \tag{2.30}$$

does not vanish,

$$\det(L) \neq 0. \tag{2.31}$$

Note that (2.31) always occurs for L being a knot. We have from (2.12), (2.13) and (2.14)

$$\Delta_K(-1) = \nabla_K(2i) \equiv 1 \bmod 4. \tag{2.32}$$

In particular, $\det(K)$ is always odd. Thus σ takes only even values on knots.

Most of the early work on the signature was done by Murasugi [Mu5], who showed several properties of this invariant.

For links $L_{\pm,0}$ with diagrams as in (2.9), we have

$$\sigma(L_+) - \sigma(L_-) \in \{0,1,2\} \tag{2.33}$$

$$\sigma(L_\pm) - \sigma(L_0) \in \{-1,0,1\}. \tag{2.34}$$

(Note: In the first property one can also have $\{0,-1,-2\}$ instead of $\{0,1,2\}$ because other authors, like Murasugi, take σ to be with opposite sign. Thus (2.33) not only defines a property, but it also specifies our sign convention for σ.)

Next, σ obeys additivity under split union (denoted by '$\sqcup$') and connected sum (denoted by '#'):

$$\sigma(L_1 \# L_2) = \sigma(L_1 \sqcup L_2) = \sigma(L_1) + \sigma(L_2). \tag{2.35}$$

The following other property is very useful: $\sigma(!L) = -\sigma(L)$, where $!L$ is the mirror image of L.

Further, Murasugi found the following important relation between $\sigma(K)$ and $\det(K)$ for a *knot* K.

$$\begin{aligned} \sigma(K) \equiv 0\,(4) \quad &\Longleftrightarrow \quad \det(K) \equiv 1\,(4) \\ \sigma(K) \equiv 2\,(4) \quad &\Longleftrightarrow \quad \det(K) \equiv 3\,(4) \end{aligned} \tag{2.36}$$

Furthermore,

$$\text{if } \det = 1, \text{ then even } 8 \mid \sigma \tag{2.37}$$

because of the property of signatures of unimodular quadratic forms.

Because of (2.32), one can modify (2.36) to

$$\begin{aligned} \sigma(K) \equiv 0\,(4) \quad &\Longleftrightarrow \quad \Delta_K(-1) > 0 \\ \sigma(K) \equiv 2\,(4) \quad &\Longleftrightarrow \quad \Delta_K(-1) < 0. \end{aligned} \tag{2.38}$$

Note further that for an *alternating* knot K of genus $g = g(K)$, because of (2.17), we have

$$\max\mathrm{cf}\,\Delta_K = \max\mathrm{cf}\,\nabla_K = [\Delta_K]_g = (-1)^g[\nabla_K]_{2g}$$

and because of Theorem 2.8.4,

$$\mathrm{sgn}(\Delta_K(-1)) = (-1)^g \mathrm{sgn}([\Delta_K]_g).$$

In combination with (2.38), this means that g and $\sigma \bmod 4$ determine the sign of $\max \mathrm{cf} \nabla_K = [\Delta]_{\pm g}$ (when Δ is normalized as in (2.16)):

$$\mathrm{sgn}([\Delta]_{\pm g}) = (-1)^{g+\sigma/2}. \tag{2.39}$$

We will return to this property toward the end of our long treatise.

A further use of (2.36) is that these conditions, together with the initial value

$$\sigma(\bigcirc) = 0$$

for the unknot and shortcuts like (2.35), allow one to calculate σ for most links, incl. all knots.

If K is a positive knot, then it was widely observed (for the first time probably in [CG]) that $\sigma(K) > 0$ (and hence $\sigma(K) \geq 2$). Przytycki noticed that signatures of positive links can be studied based on Taniyama's previous work [Ta], but their manuscript [PT], despite being announced long ago, was only recently completed. In particular, $\sigma(K) \geq 4$ if $g(K) \geq 2$; also $\sigma(K) > 0$ if K is almost positive or 2-almost positive, except a twist knot. See [St3, St7, St2] for previously published (though not previously written, and not identical) proofs (and Chapter 6 below for more discussion).

There is a nice combinatorial formula for σ for alternating diagrams. We use this formula here to remind us of the below property of signatures for alternating links.

Proposition 2.10.1 Let L be a non-split alternating link. Then

$$\sigma(L) \leq 1 - \chi(L), \tag{2.40}$$

and equality holds if and only if L is special alternating and positive.

Here we emphasize the equality case far more than (2.40) itself. This inequality is true for much more general links, including all links with connected minimal genus Seifert surfaces, in particular also all non-split positive links and links with non-zero Alexander polynomial (recall Theorem 2.8.1). For knots (where always $\Delta \neq 0$), (2.40) follows as a combination of (2.15) and Theorem 9.1.3. (There are ways to extend this reasoning to links, but they definitely lead too far afield here.)

Proposition 2.10.1 is well known but will be relevant in the following, and because I am not certain where to find a proof, I include one.

Proof. This inequality can be proved using checkerboard colorings (§2.5.2). We consider now only the case that D is alternating. Then it is well known that all A-corners of all crossings of D according to Figure 2.9 arise as corners of regions of one color, and all B-corners for the regions of the other color. Our convention is that the white regions contain all A-corners.

By Corollary 2.8.3, D is non-split. By Corollary 2.8.7, we need to show that equality in (2.40) occurs if and only if D is positive.

We will use the formula for σ on p. 437 of [Kf] and negate the r.h.s. (we want the positive trefoil to have positive signature; the convention there is opposite). We get, using (2.7),

$$\sigma(D) = \frac{w(D)}{2} + \frac{c(D) - 2(W(D) - 1)}{2}. \tag{2.41}$$

Then with $w(D) = c(D) - 2c_-(D)$, (2.40) transcribes into

$$W(D) \le s(D) + c_-(D). \tag{2.42}$$

We will see this inequality (together with its equality case) by induction on the number of blocks or Murasugi atoms (see definition in §2.9).

To start induction, assume D is special. Since D is non-split and alternating, we know that D is positive or negative. If D is positive, then $c_-(D) = 0$, and $W(D) = s(D)$, and we have (2.42) with equality. If D is negative, then $B(D) = s(D)$ and $c_-(D) = c(D)$. Thus (2.42) becomes, in a strict form,

$$c(D) + 2 - s(D) < c(D) + s(D),$$

which is true for $s(D) > 1$. (When $s(D) = 1$, then D is a circle.)

To complete induction, it suffices to observe how the quantities in (2.42) behave under Murasugi sum $*$. We have for $D = D_1 * D_2$,

$$\begin{aligned} W(D) &= W(D_1) + W(D_2) - 1, \\ s(D) &= s(D_1) + s(D_2) - 1, \\ c_-(D) &= c_-(D_1) + c_-(D_2). \end{aligned}$$

For the first equality, note that under Murasugi sum, one region of D_1 and one region of D_2 are annihilated; and exactly one is black and one is white. This shows (2.42) (including its equality case) and proves Proposition 2.10.1. □

2.11 Genus generators

Now let us recall, from [St4, St2], some basic facts concerning knot generators of a given genus. An explanation is also given in section 5.3 in [Cr2]. (There are several equivalent forms of these definitions, and we choose here one that closely leans on the terminology of Gauss diagrams; see for example [St].) We will also set up some notations and conventions used below. The situation for links is discussed only briefly here and in much more detail in Chapter 3.

Definition 2.11.1 We call two crossings p, q in a knot diagram *linked*, and write $p \cap q$ if passing their crossing points along the orientation of D we have the cyclic order $pqpq$ and not $ppqq$.

Definition 2.11.2 Let D be a knot diagram, and p and q be crossings.

(i) We call p and q *(twist) equivalent*, $q \simeq p$, if for all $r \neq p,q$ we have $r \cap p \iff r \cap q$.

(ii) We call p and q *$\sim$-equivalent* and write $p \sim q$ if p and q are equivalent and $p \not\cap q$.

(iii) Similarly, p and q are called *$\overset{\sim}{*}$-equivalent*, $p \mathrel{\overset{\sim}{*}} q$, if p and q are equivalent and $p \cap q$.

(iv) Finally, call two crossings p and q *Seifert equivalent* if they connect the same two Seifert circles.

Let us record the following easy but useful observation.

Lemma 2.11.3 $\overset{\sim}{*}$-equivalent crossings are Seifert equivalent. The converse is true in special diagrams. □

Definition 2.11.4 A $\sim$-equivalence class consisting of one crossing is called *trivial* and a class of more than one crossing *non-trivial*. A $\sim$-equivalence class is *reduced* if it has at most two crossings; otherwise it is *non-reduced*.

Definition 2.11.5 Let $t(D)$ be the number of $\sim$-equivalence classes of D. For $i = 1,\dots,t(D)$ let $t_i(D)$ be the number of crossings in the i-th $\sim$-equivalence class. Then for $i = 1,\dots,t(D)$ and $j = 1,\dots,t_i(D)$ let $p(D,i,j)$ be the j-th crossing in the i-th $\sim$-equivalence class of D.

Definition 2.11.6 A $\bar{t}'_2$ *move* or *twist* is given up to mirroring by

(2.43)

We will call it positive or negative, depending on whether it acts on a crossing of the according sign.

Thus a $\bar{t}'_2$ move at a crossing x in a diagram D creates a pair of $\sim$-equivalent crossings to x. (This property specifies the $\bar{t}'_2$ move alternatively up to flypes.)

Definition 2.11.7 An alternating diagram D is called *$\bar{t}'_2$ irreducible* or *generating diagram* if all $\sim$-equivalence classes are reduced, that is, $t_i(D) \leq 2$ for $i = 1,\dots,t(D)$. An alternating knot K is called a *generator* if some of its alternating diagrams is generating. The diagrams obtained from D by $\bar{t}'_2$ moves and crossing changes form the *sequence* or *series* $\langle D \rangle$ of D.

(In many cases it will be useful to restrict oneself in $\langle D\rangle$ to alternating or positive diagrams.)

Observe that, because $\sim$-equivalence is invariant under flypes, Theorem 2.6.2 implies that some alternating diagram of K is generating if and only if all its alternating diagrams are such.

A flype in Figure 1.2 is called *trivial* if one of the tangles P, Q contains only crossings equivalent to the crossing admitting the flype. A flype is of type A if the strand orientation is so that strands on the left/right side of each tangle are directed equally w.r.t. the tangle (i.e., both enter or both exit). Otherwise it is a flype of type B. Compare Figure 2.6. So the property of a crossing to admit a type A respectively type B flype is an invariant of the $\underset{*}{\sim}$ respectively $\sim$-equivalence class.

Theorem 2.11.8 ([St4]) There exist only finitely many generators of a given genus. All diagrams of that genus can be obtained from diagrams of these generators under $\bar{t}'_2$ twists, flypes, and crossing changes.

It may be worth noticing that in this sequence of operations, all flypes can be performed at the beginning.

The finiteness of generators, together with the Flyping theorem [MT], shows the following theorem.

Theorem 2.11.9 (see [St4]) Let $a_{n,g}$ be the number of prime alternating knots K of genus $g(K)=g$ and crossing number $c(K)=n$. Then for $g\geq 1$

$$\sum_n a_{n,g}x^n = \frac{R_g(x)}{(x^{p_g}-1)^{d_g}},$$

for some polynomial $R_g\in\mathbb{Q}[x]$, and $p_g, d_g\in\mathbb{N}$.

Alternatively, this statement can also be written in the following form: There are numbers p_g (period), n_g (initial number of exceptions) and polynomials $P_{g,1},\dots,P_{g,p_g}\in\mathbb{Q}[n]$ with

$$a_{n,g} = P_{g,n \bmod p_g}(n) \text{ for } n\geq n_g\,. \tag{2.44}$$

This was explained roughly in [St4] and then in more detail in [SV], where we made effort to characterize the leading coefficient of these polynomials $P_{g,i}$. Even if the polynomials vary with a very large period p_g, the leading coefficients depend only on the parity of n (see §10.3). The degrees of all $P_{g,i}$ are also the same and equal to 1 less than the maximal number of $\sim$-equivalence classes of diagrams of canonical genus g, with the exception $g=1$, in which case this degree is 1 or 2 depending on whether i is even or odd.

In practice (in particular as we will see below), it is important to obtain the list of generators for small genus. Genus 1 is easy and was also observed independently.

Theorem 2.11.10 ([St4]; see also [Ru]) There are two generators of genus 1, the trefoil and figure-8-knot.

Genus 2 and 3 require much more work. For reasons already suggested, it is sufficient to find prime diagrams, and by Theorem 2.1.1, prime generators.

Theorem 2.11.11 ([St2]) There are 24 prime generators of genus 2, which are 5_1, 6_2, 6_3, 7_5, 7_6, 7_7, 8_{12}, 8_{14}, 8_{15}, 9_{23}, 9_{25}, 9_{38}, 9_{39}, 9_{41}, 10_{58}, 10_{97}, 10_{101}, 10_{120}, 11_{123}, 11_{148}, 11_{329}, 12_{1097}, 12_{1202}, and 13_{4233}, and 4017 prime generators of genus 3.

A classification (by means of obtaining the list of prime generators) for genus $g=4$ is also possible and will be explained below, along with a general statement in Theorem 3.1.1.

It follows from Theorem 2.6.2 that the series of different (alternating) diagrams of the same generating knot are equivalent *up to mutations*. These mutations arise when a $\bar{t}'_2$ move is performed at a crossing admitting a type A flype. When using tests involving the polynomial invariants, which do not change under mutations, it is legitimate that *a priori* we fix *a single specific diagram D for each generator K* and work only with (the series of) this diagram. Then we write $t(K)=t(D)$, $t_i(K)=t_i(D)$ and $p(K,i,j)=p(D,i,j)$ and speak of the sequence or series of K. With some simple observations, we will be able to comfort ourselves with this assumption in Chapter 7, whereas in Chapter 5 more care is needed.

Further, we will often discard diagrams with crossings of different sign within the same $\sim$-equivalence class. Such diagrams have a trivial clasp after flypes and are usually of little interest. We will call a $\sim$-equivalence class *positive* or *negative* depending on the sign of its crossings. There is an important exception to this rule, which occurs for $t_i=2$: Then we do not cancel one positive and one negative crossing (because we need to keep the $\sim$-equivalence class).

Turning to *link* diagrams, analogous, although slightly modified, definitions are possible. We call alternating diagrams *generating* or *$\bar{t}'_2$-irreducible* if they cannot be reduced by flypes and inverses of the move (2.43). Definition 2.11.7 of a *series* can be restated here.

The finiteness result in Theorem 2.11.8 can be extended to link diagrams of a given genus and a given number of components. We will discuss this in detail in Theorem 3.1.1.

For link diagrams, we can define two crossings as *$\sim$-equivalent* if after a sequence of flypes they can be made to form a *reverse clasp*  ; it is an exercise to check that this is an equivalence relation (see [St2]). Similarly, we call two crossings *$\underset{*}{\sim}$-equivalent* if after a sequence of flypes they can be made to form a *parallel clasp*  . We observed (see [St4, St2, STV]) that $\sim$- and $\underset{*}{\sim}$-equivalent crossings of a knot diagram are linked with the same set of other crossings in the diagram.

Thus, the present definition is a genuine extension of the previous one for the case of knots.

In relation to clasps, let us stipulate that the clasp is *positive*, *negative*, or *trivial/resolved*, depending on the sign of its crossings. Removing a trivial clasp by a Reidemeister II move, after possibly switching a crossing in a non-trivial clasp, will be called *resolving* the clasp.

2.12 Knots vs. links

To conclude our preparation, let us make more precise the separation between knots and links in the following work.

The methods developed in this text put into prospect to adapt results about "k-almost positive knots" and "alternating knots of genus k" to statements that apply to "$\tilde{k}$-almost positive l-component links" and "alternating l-component links of genus $\tilde{k}$" for $\tilde{k}+l \leq k+1$.

One main technical reason for focusing on knots is that our computational tools are heavily based on knot diagrams (see similarly Remark 5.4.5). The results where links are covered appear in Chapter 3 (here the inclusion of links is essential and will be used elsewhere) and in §7.5. They rely on purely theoretical arguments, and computation is not needed. To some extent, we managed to adapt computations to the link case, as shown in Theorem 9.2.7. Its proof should serve to explain how to extend arguments to links in other parts as well, however, more effort will be needed. We also expect that some statements for links would look less pleasant than for knots.

Chapter 3

The maximal number of generator crossings and ∼-equivalence classes

3.1 Generator crossing number inequalities

As explained in the introduction, we must first spend some effort in deriving estimates for the crossing number of generators of a given canonical Euler characteristic. These estimates turn out to be rather sharp and are a consequence of a detailed study of the special diagram algorithm of Hirasawa [Hi2] and myself [St9, Section 7].

Theorem 3.1.1 In a connected link diagram D of canonical Euler characteristic $\chi(D) \le 0$, there are at most

$$t(D) \le \begin{cases} -3\chi(D) & \text{if } \chi(D) < 0 \\ 1 & \text{if } \chi(D) = 0 \end{cases} \tag{3.1}$$

∼-equivalence classes of crossings. If D is $\bar{t}_2'$-irreducible and has $n(D)$ link components, then

$$c(D) \le \begin{cases} 4 & \text{if } \chi(D) = -1 \text{ and } n(D) = 1\,, \\ 2 & \text{if } \chi(D) = 0\,, \\ -6\chi(D) & \text{if } \chi(D) < 0 \text{ and } n(D) = 2 - \chi(D)\,, \\ -5\chi(D) + n(D) - 3 & \text{else}\,. \end{cases} \tag{3.2}$$

We will show below (Proposition 3.4.1) that all inequalities (3.2) are the best possible.

The inequality (3.1) is also optimal, and we call the generating diagrams D that make (3.1) exact *maximal generating diagrams* and their knots/links *maximal generators*. For knots, these generators were studied in detail in [SV].

The main idea behind the description of diagrams of a given canonical Euler characteristic was to show that they decompose into finitely many equivalence classes under $\bar{t}'_2$ twists and their inverses. In [St4], we showed this only for knots, but the case of links can be easily recurred to it. For a diagram D of a link L with $n(L)$ components, one applies $n-1 = n(L)-1$ moves, which replace a positive/negative crossing of two different components by a parallel positive/negative clasp:

(3.3)

We call this procedure a *clasping*. Thus, D can be transformed into a knot diagram D' with $1-\chi(D') = n-\chi(D)$.

Although this simple argument establishes the picture qualitatively, it is not useful for an optimal estimate and neither was our original approach in [St4] for knots. Then, in subsequent work in [STV, SV], we established a relation between canonical Seifert surfaces for knot diagrams and 1-vertex triangulations of surfaces and obtained some partial information (certainly much better than in [St4]) on the maximal number of crossings and $\sim$-equivalence classes of $\bar{t}'_2$-irreducible knot diagrams of a given canonical genus. Unfortunately, this method does not extend pleasantly to links. We will therefore now introduce a different approach, which is entirely knot theoretic and circumvents this problem. (Contrarily, there are insights gained from the old approach, which will also be used later but which cannot be recovered.) This approach originates from an algorithm, first found by Hirasawa [Hi], to make any link diagram into a special one without altering the canonical Euler characteristic (even, in fact, its isotopy type).

We will apply this algorithm to prove the above result given in Theorem 3.1.1, which is likely to be in its optimal and most general form. Before the proof of Theorem 3.1.1, we will first discuss Hirasawa's algorithm.

3.2 An algorithm for special diagrams

We recall the following terminology (§2.9).

Definition 3.2.1 Seifert circles in an arbitrary diagram are called *non-separating*, if they have empty interior or exterior; the others are called *separating*. A diagram is called *special* if all its Seifert circles are non-separating. Any link diagram decomposes along its separating Seifert circles as the *Murasugi sum* ($*$-product) of special diagrams (see [Cr, Section 1]).

In [BZ], it was proved that each link has a special diagram by a procedure to turn any given diagram of the link into a special one. However, the procedure in this

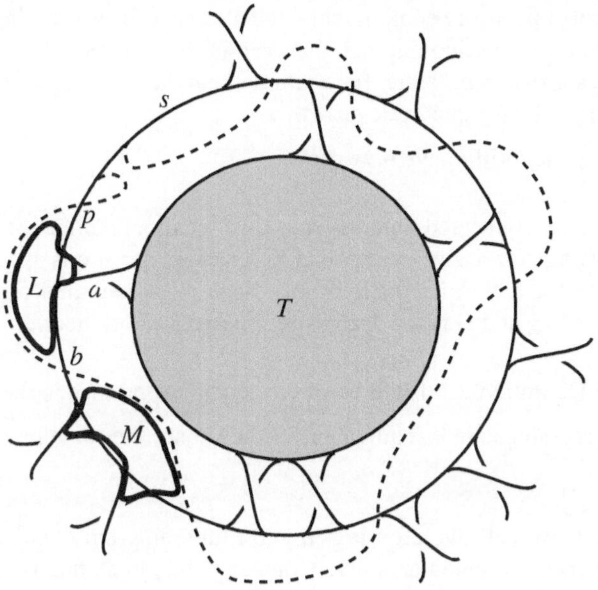

FIGURE 3.1: Seifert circles near a Hirasawa move I.

proof drastically alters the initial diagram and offers no reasonable control on the complexity (canonical genus and crossing number) of the resulting special diagram. A much more economical procedure was given by Hirasawa [Hi] and independently rediscovered a little later in [St9, Section 7]. Hirasawa's move consists of laying a part of a separating Seifert circle s along itself in the opposite direction (we call this move *rerouting*; it is the opposite of the wave move in Figure 2.7), while changing the side of s depending on whether interior or exterior adjacent crossings to s are passed. See Figure 3.1.

The move of [St9] is similar, except that this type of rerouting is applied to the Seifert circle connected to s by an crossing c exterior to s. This move lowers the canonical Euler characteristic by two, but by properly choosing to reroute the strand above or below the rest of the diagram (that is, such that it passes newly created crossings all as over- or all as undercrossings) one obtains a trivial parallel clasp involving c, whose deletion raises the canonical Euler characteristic back by two. Then we obtain an instance of Hirasawa's move.

Hereby, unlike in Hirasawa's original version of his algorithm, we take the freedom to alter the signs of the new crossings, as far as the isotopy type of the link is preserved, but it is *not* necessarily to preserve the canonical Seifert surface. It is of importance to us only that the canonical Euler characteristic of the diagram is preserved. We assume that this freedom is given throughout the rest of this chapter.

Hirasawa's algorithm is very economical—the number of new crossings added is linearly bounded in the crossing number and even in the canonical Euler characteristic of the diagram started with. (Note, for example, that the braid algorithms of Yamada [Ya] and Vogel [Vo] have quadratic growth.)

Using Hirasawa's algorithm, we now prove:

Lemma 3.2.2 Any connected link diagram D of maximal number of $\sim$-equivalence classes of crossings for a given canonical Euler characteristic is special. Moreover, this is also true for any $\bar{t}'_2$-irreducible link diagram of maximal crossing number, except if $\chi(D) = -1$ and D is the 4-crossing (figure-8-knot) diagram.

The notions in Definition 2.5.1 will be of particular importance in the proof.

Note that two crossings are $\sim$-equivalent iff they share the same pair of non-Seifert circle regions.

Definition 3.2.3 We call two crossings a and b in a diagram D *neighbored*, if they belong to a reversely oriented primitive Conway tangle in D, that is, there are crossings $c_1, \ldots, c_n$ with $a = c_1$ and $b = c_n$, such that c_i and c_{i+1} form a reverse clasp in D.

This is a similar definition to $\sim$-equivalence but with no flypes allowed. Thus, the number of $\sim$-equivalence classes of a diagram is not more than the number of neighbored equivalence classes of the same diagram or of any flyped version of it.

Proof of Lemma 3.2.2. We will first consider the second statement, that is, the one for the crossing number of $\bar{t}'_2$-irreducible link diagrams.

To show this, take a non-special reduced diagram D (not necessarily a generator). Apply flypes so that all $\sim$-equivalent crossings are neighbored equivalent. (This is possible because flyping in a flyping circuit is independent from the other ones. For the definition of flyping circuits and related discussion see [ST, Section 3]. Alternatively, one can consult Lackenby's paper [La] where such diagrams are called "twist reduced.")

We will now show that the application of Hirasawa moves and appropriate flypes augments the crossing number of the generator, in whose series the diagram lies, and preserves the condition that $\sim$-equivalent crossings are neighbored equivalent. The flypes we will apply always reduce the number of separating Seifert circles, and so does any Hirasawa move. Thus, we will have done our job by induction on the number of separating Seifert circles.

Clearly, one can assume that D is prime. The composite case follows easily from the prime one.

Consider again the picture of the Hirasawa move, Figure 3.2. Here we avoid the creation of the nugatory crossing p in Figure 3.1. Let s be the separating Seifert circle of D on which the move is performed.

We distinguish three cases according to $\operatorname{ind}(s)$.

Case 1. If $\operatorname{ind}(s) = 1$, D is not prime.

Case 2. Now assume that $\operatorname{ind}(s) \geq 3$. Then s has $\operatorname{ind}(s)$ adjacent regions from inside and outside. In Figure 3.2, the inner regions are called A, A_1, A_2 and A_3, and the outer regions are called B, B_1, B_2 and B_3. The Hirasawa move splits up from every such region R a small part R', containing a new Seifert circle. (In Figure 3.2, four of the 8 such R' are displayed.) Then it joins an inner and outer region (here A and B) to a new region we call AB. It adds $2\operatorname{ind}(s) - 1$ crossings. We call these crossings *new*, the others, existing already before the move, *old*. Call the new diagram D'.

We show that D' is reduced. Assume that p is a nugatory crossing in D'. Then its two non-Seifert circle regions coincide. If p is an old crossing, the only way to make them distinct by undoing the Hirasawa move is if they are A and B. However, the non-Seifert circle regions of any crossing in D lie either both inside or both outside of s. If p is new, then we directly see that the non-Seifert circle regions lie one inside and one outside of s and, hence, are also distinct.

Now we examine how the $\sim$-equivalence relation has been altered under the Hirasawa move.

Assume $p \sim q$ in D'. Then p and q have the same pair of non-Seifert circle regions. (Because no crossing in D' is nugatory, the two non-Seifert circle regions are always distinct.)

1) Assume first p and q are old crossings in D' and $p \sim q$. Then the non-Seifert circle regions at p and q have remained the same when undoing the Hirasawa move (possibly the small Seifert circle parts are removed), except that the region AB has been separated. Thus $p \sim q$ in D also, except if there is a region C in D such that the non-Seifert circle regions at p in D are A and C, and the non-Seifert circle regions at q are B and C. However, the non-Seifert circle regions of any crossing lie either both inside or both outside of s. Because A and B lie on different sides of s, C cannot exist. Thus, $p \sim q$ in D.

2) If p and q are both new, then the non-Seifert circle regions can be explicitly given among the inner and outer adjacent regions of s (in our example the As and Bs). It is easy to see that from the definition of the Hirasawa moves, all pairs (A_i, B_i) and (A_i, B_{i+1}) occurring this way are distinct. (Here one needs that D is prime; some of the A_i or B_i alone may coincide!) Thus, $p \not\sim q$.

3) Finally, consider the case p new, q old. The two non-Seifert circle regions of p lie one in- and one outside of s in D. By the argument with the region C in Case 1), we have $p \sim q$ only if p and q are adjacent to the new region AB in D'. From the description of pairs of non-Seifert circle regions of the new crossings and primeness of D, as mentioned in Case 2), there are two such crossings p. They are the first and last on the newly created strand. They are named p_1 and p_2 in Figure 3.2. They are unique because of the argument in Case 2).

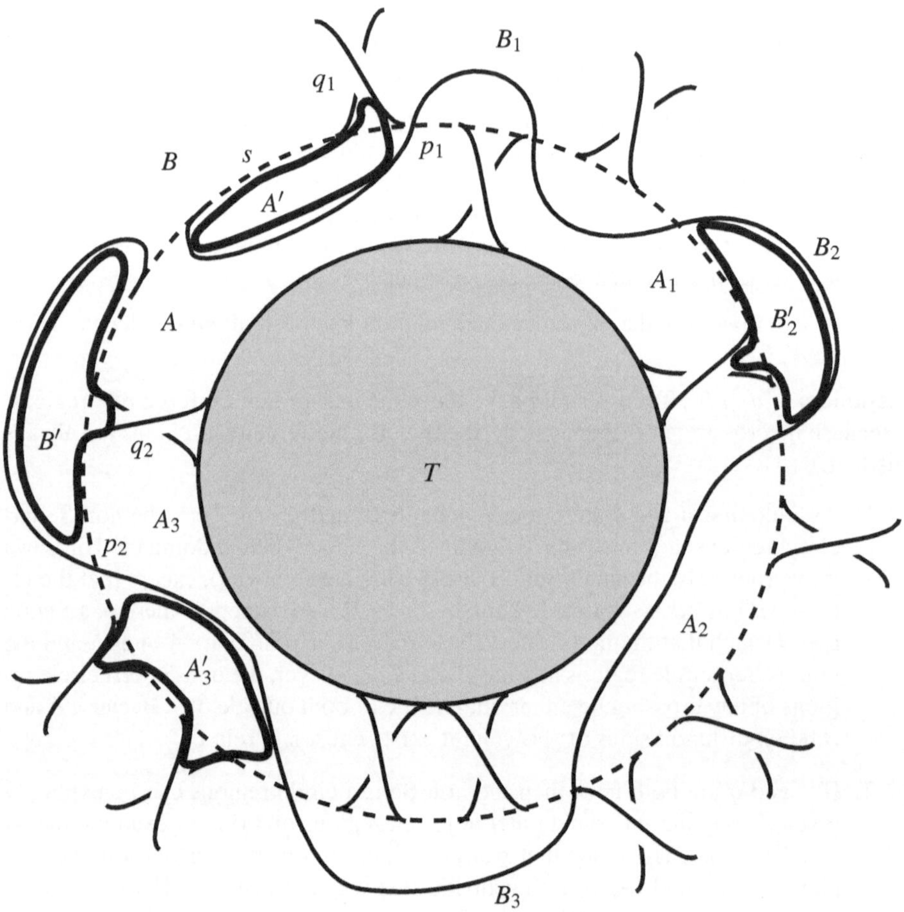

FIGURE 3.2: Seifert circles near a Hirasawa move II.

We proved that the only possible newly created pairs of $\sim$-equivalent crossings are (p_1, q_1) and (p_2, q_2), in which q_1 nd q_2 are taken from two fixed (distinct) $\sim$-equivalence classes in D.

The addition of a crossing to a $\sim$-equivalence class can reduce the crossing number of the underlying generator by at most 2 (which occurs if the crossing added is of odd number in its $\sim$-equivalence class). Because $2\,\mathrm{ind}(s) - 1 \geq 5$ new crossings are added, it follows that the crossing number of the generator goes up under a Hirasawa move.

Case 3. Finally, consider the slightly tricky case $\mathrm{ind}(s) = 2$. Now $2\,\mathrm{ind}(s) - 5 < 0$ so that we cannot argue as before. Let $n_i(s)$ and $n_o(s)$ be the number of inner and outer crossings adjacent to s respectively. Clearly $n_i(s), n_o(s) \geq \mathrm{ind}(s) = 2$. If $n_i(s) = n_o(s) = 2$, the two internal and external crossings of s are $\sim$-equivalent. Then by assumption they are also neighbored equivalent, and then we have the figure-8-knot diagram of genus one, which we excluded. Thus, assume w.l.o.g. (up to S^2-moves) that $n_i(s) \geq 3$. Then we apply a *modified* Hirasawa index-2-move:

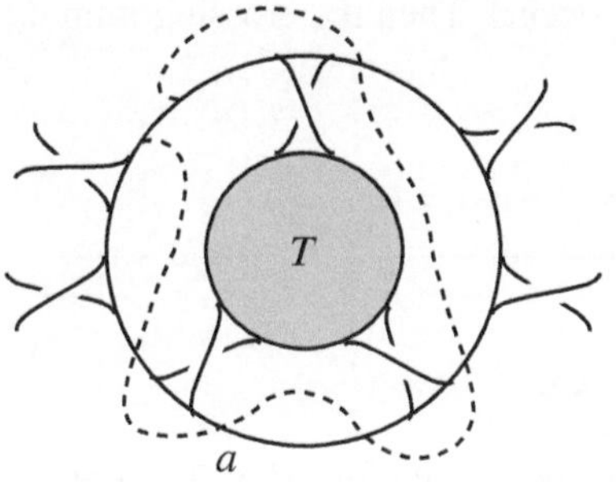

(3.4)

This means we create an additional trivial clasp in one of the inner groups of at least 2 crossings.

It is easy to see that this move augments the number of generator crossings (at least) by 1.

The condition that $\sim$-equivalent crossings are neighbored equivalent is not necessarily preserved under the Hirasawa move. However, using flypes (which preserve the number of crossings and $\sim$-equivalence classes), one can again reestablish it. It is an easy observation that the needed flypes reduce the number of separating Seifert circles.

Then iterate Hirasawa moves (and possible flypes) until D has become special. Then we know, by this argument, that the crossing number of the generator of this special diagram is strictly greater than this of D. This proves the second assertion in the lemma.

The proof of the first part of the assertion goes along similar lines, but it is simpler. Because we showed that the number of $\sim$-equivalence classes raises by $2\,\mathrm{ind}(s) - 3$

under each Hirasawa move, it is sufficient to assume $\operatorname{ind}(s) \geq 2$. This makes the flyping argument unnecessary, the need to care about $\sim$-equivalent crossings being neighbored equivalent, or about $\operatorname{ind}(s) = 2$, and eliminates the exception of the 4-crossing diagram. □

Corollary 3.2.4 For given n and χ, except $(n,\chi) = (1,-1)$, a prime generator diagram D of maximal crossing number is special and has maximal number of $\sim$-equivalence classes.

Proof. That D is special follows from Lemma 3.2.2. If D is special and does not have the maximal number of $\sim$-equivalence classes, then by the work of [SV] we know that its unbisected Seifert graph G' is either not trivalent or not 3-connected. (The argument there was applied for knots, but this condition was not used.) The latter case reduces to former, because if G' is not 3-connected then one can apply a flype on D to have a vertex of valence ≥ 4 in G'. In [SV], we argued (under the exclusion of cut vertices) that one can apply a decontraction on such a vertex, such that the created clasp is a new $\sim$-equivalence class (and among the previous $\sim$-equivalence classes no identifications occur). Then the crossing number of the generator is also augmented. □

3.3 Proof of the inequalities

The lemma does the main part of the proof of Theorem 3.1.1, which will now be completed. (Some of the arguments that follow also appear in very similar form in [SV], so that the reader may consult there for more details.)

Proof of Theorem 3.1.1. The lemma shows that for obtaining the stated estimates, only special diagrams need to be considered. (Because the 4-crossing diagram does not violate any of the two assertions of the theorem, let us exclude it in any further consideration.)

Assume that D has a Seifert circle of valence ≥ 4. If we have a $\bar{t}'_2$-irreducible special diagram with a Seifert circle of valence ≥ 4, then one can always perform a Reidemeister II move on any pair of non-neighbored edges in this Seifert circle region. Then one obtains a special diagram of the same Euler characteristic and two more crossings. By proper choice of the pair of edges, the new crossings can be made to form a separate new $\sim$-equivalence class so that the diagram is still $\bar{t}'_2$-irreducible (although not always any pair of such edges will do).

This means that we need to consider only special diagrams D with Seifert circles of valence 2 and 3. In other words, the Seifert graph (see [Cr, Section 1]) G of D is 2-3-valent (all its vertices have valence 2 or 3). Removing vertices of valence 2 in G by unbisections means identifying edges that correspond to $\sim$-equivalent crossings in D. Let G' be the 3-valent (cubic) graph obtained this way. Because D is reduced, G' is connected and contains no loop edges. Note that G is a reduced bisection of G'.

However, G' may be a single loop, which occurs when $\chi(D) = 0$. Then one obtains the $(2,k)$-torus link diagrams with reverse orientation (which are generated by the Hopf link and have one ~-equivalence class). Thus, we can exclude the degenerate case $\chi(D) = 0$ in the rest of the proof.

Then, if $e(G')$ is the number of edges of G' and $v(G')$ the number of its vertices, we have $v(G') = {}^2\!/_3 e(G')$, and

$$-\frac{e(G')}{3} = v(G') - e(G') =: \chi(G') = \chi(G) = \chi(D).$$

Because the number of ~-equivalence classes in D is at most $e(G')$, we obtain the first result.

Now consider the second statement in the theorem. From Hirasawa's algorithm we know that the maximal crossing number of a generator is achieved by a special diagram (except for the case of knot diagrams of genus one), and more precisely in such one whose Seifert graph is 2-3-valent. Such a graph has $2 - \chi(D)$ faces (corresponding to regions of D that do not contain a Seifert circle). For a while, forget about the orientation of any component. If every ~-equivalence class of such a diagram D had 2 crossings, the number of components of D would be equal to its number $2 - \chi(D)$ of non-Seifert circle regions. Because the change of 2 crossings to 1 in each ~-equivalence class

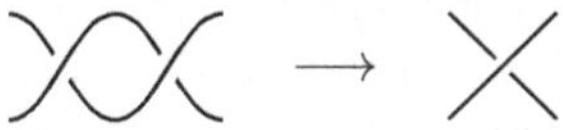 (3.5)

changes the number of components at most by ± 1, we need at least $2 - \chi(D) - n(D)$ such replacements to obtain a diagram of $n(D)$ components. Thus, the maximal number of crossings we can have is

$$-6\chi(D) - \bigl(2 - \chi(D) - n(D)\bigr) = -5\chi(D) + n(D) - 2,$$

and we showed

$$c(D) \le \begin{cases} -5\chi(D) + n(D) - 2 & \text{if } \chi(D) < 0 \\ 2 & \text{if } \chi(D) = 0 \end{cases}. \tag{3.6}$$

To obtain (3.2), we must show that we can improve the estimate in (3.6) by one crossing if $\chi < 0$ and $n < 2 - \chi$ (we excluded knot diagrams of genus one).

Assume there is a diagram D with $-5\chi + n - 2$ crossings. Let G be its 2-3-valent Seifert graph, and let G' be the 3-valent graph obtained from G by deleting valence-2 vertices. Call the edges of G' obtained under such unbisections *even*, and the others *odd*. The graph G' inherits a particular planar embedding from G and, hence, we can build its dual graph G'^*. Note that each edge e' of G' corresponds bijectively to an edge e'^* in G'^*. Define for a subgraph $\Gamma \subset G'$ the "dual" graph

$$\Gamma^* := \{e'^* : e' \in \Gamma\}.$$

Let

$$\Gamma := \{e' \in G' : e' \text{ odd}\}. \tag{3.7}$$

We claim that Γ^* is loopless. Assume that Γ^* contains a loop made up of edges $l_1,\dots,l_k$. Take the diagram D_0 with the same graph G' but such that all edges are even. Then D is obtained from D_0 by $2-n-\chi$ moves (3.5), corresponding to the odd edges of G'. (The order of these moves is irrelevant.) To have n components in D (and mot more), each such move must reduce the number of components by one, so that both strands on the left of (3.5) must belong to different components. However, if one performs on D_0 the moves (3.5) corresponding to $l_1,\dots,l_{k-1}$, and then one applies the move for l_k, it is easy to see that both strands on the left of (3.5) for this last move belong to the same component. This contradiction shows that Γ^* is loopless.

Because Γ^* is loopless, $\Gamma^* \subset \Gamma'^*$ for some spanning tree Γ'^* of G'^*. Then $G' \setminus \Gamma'$ is also a spanning tree (of G'). Thus, G' has a spanning tree made up of even edges.

To show that this is impossible, consider the orientation of the Seifert circles corresponding to the vertices of G. Sign the vertices of G positive or negative depending on this orientation. (This makes G bipartite.) Then this signing also reduces to a signing of G', with the property that even edges connect vertices of the same sign and odd edges connect vertices of opposite sign. Because $n < 2-\chi$, there must be vertices of G' of both signs. But G' has a spanning tree made up of even edges, and so all of its vertices must have the same sign, which is a contradiction. (Clearly, adding just one edge to Γ may solve the problem because Γ^* may no longer be loopless and $G' \setminus \Gamma$ may get disconnected.)

This completes the proof of Theorem 3.1.1. □

Remark 3.3.1 Note that we always have $2-\chi(D) \geq n(D)$, so that we can always eliminate n for χ.

Remark 3.3.2 In practice, it will often be convenient to use (3.6) rather than (3.2) because we can absorb by this additional crossing two of the exceptional cases (and avoid unpleasant case distinctions). Nonetheless, the additional argument for (3.2) is useful to completely settle the question on the maximal crossing numbers of generators in many cases, in particular for knots.

Remark 3.3.3 It is clear from the proof that there is no natural way to obtain an upper restriction on the number of $\sim$-equivalence classes of odd number of crossings. In fact, no such restriction exists: In the case of knots and genus $g \geq 6$, we constructed in [St2] generator diagrams with the maximal number of $-3\chi = 6g-3$ $\sim$-equivalence classes, each one having a single crossing.

Remark 3.3.4 If G' is 3-connected, then it is easy to see that in fact different edges of G' correspond to different $\sim$-equivalence classes in D, so that one easily finds examples where the first bound -3χ in Theorem 3.1.1 is realized sharply. We will study such generators D in more detail in a subsequent paper [St10].

3.4 Applications and improvements

We start with some promised examples showing that for a general generator, the estimates (3.2) are the best possible.

Proposition 3.4.1 There are $\bar{t}'_2$-irreducible diagrams D for which all inequalities (3.2) are exact, i.e., equalities.

Proof. Examples showing the sharpness of the first three (exceptional) cases in (3.2) follow directly from the proof of Theorem 3.1.1. Thus we explain only the (main) fourth case.

The end of the proof of Theorem 3.1.1 suggests how to find such examples. First, they must be maximal, and thus their unbisected Seifert graph G' is 3-valent. Moreover, the dual edges Γ^* of Γ (the odd edge subgraph) in (3.7) must form exactly one cycle. With this insight, giving the desired examples is easy.

We consider the following 3-valent graphs G' with distinguished vertices $c_1, \dots, c_k$ (for $k \geq 3$).

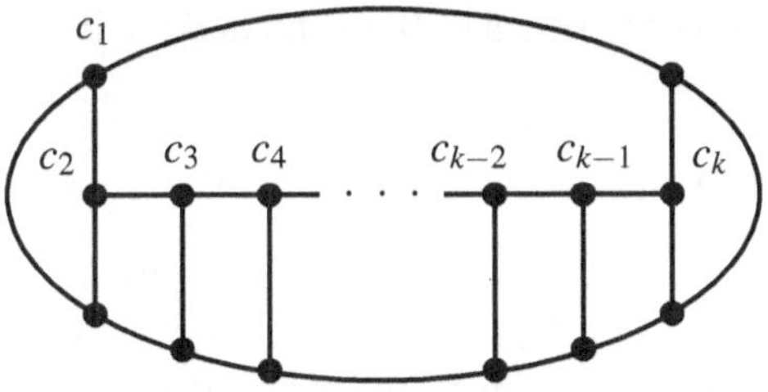

To determine D, it is enough to specify the orientation ($+$ or $-$) or the 3-valent Seifert circles. (For $\chi = -2$, labeling any two vertices of the tetrahedral graph with either sign will do.)

First, make all c_i positive and all other vertices negative. It is easy to check that this gives examples of D of the desired type with $n(D) = 1$ for k odd (a series which was already noticed in [SV]) and $n(D) = 2$ for k even.

To obtain examples for higher $n(D)$, successively turn negative the sign of $c_k, c_{k-1}, \dots$ (while keeping the non-labeled vertices negative). □

Thus, we settle the problem of how to determine the maximal crossing number of a generator for knots.

Corollary 3.4.2 The maximal crossing number of a knot generator of genus $g \geq 2$ is $10g-7$. □

Now we give some first applications of the crossing number estimates for generators. For this we use the degree-2-Vassiliev knot invariant

$$v_2(K) = 1/2\Delta_K''(1) = [\nabla_K]_2 .$$

Beside the independent interest of these inequalities, their proofs also introduce an idea that, in modified form, will be important for later more general results, in particular for the work in [St8].

Theorem 3.4.3 If K is a positive knot, then each reduced positive diagram D of K has $c(D) \leq 9g(K)-8+2v_2(K)$ crossings, except if K is the trefoil (where $c(D) \leq 4$).

Proof. Let D' be the positive generator in whose series D lies. We have $g(D') = g(D) = g(K)$, and we know that a positive $\bar{t}_2'$-twist in a positive diagram augments v_2. If $g(D') > 1$, then $c(D') \leq 10g(D')-7$ by Corollary 3.4.2. On the opposite side, we have $c(D') \geq 2g(D')+1$, so that by theorem 6.1 of [St]

$$v_2(D') \geq \frac{c(D')}{4} \geq \frac{g(D')+1}{2}$$

(where in the second inequality we used the obvious integrality of v_2). Now we can apply at most $v(D) - v_2(D')$ $\bar{t}_2'$-twists to D' to obtain D, so that

$$\begin{aligned} c(D) &\leq c(D')+2\big(v_2(K)-v_2(D')\big) \\ &\leq 10g(D')-7+2v_2(K)-g(D')-1 = 9g(K)-8+2v_2(K). \end{aligned}$$

If $g(D) = g(D') = 1$, then we must show $c(D) \leq 2v_2(K)+1$. This can be checked directly, because by Theorems 2.11.10 and 2.11.11, D is either a (p,q,r)-pretzel diagram for $p,q,r \geq 1$ odd, or a positive rational knot diagram $C(p,-q)$, with $p,q \geq 2$ even. In the former case $c(D) = p+q+r$ and $v_2(D) = (pq+pr+qr+1)/4$, and the inequality easily follows. In the latter case $c(D) = p+q$ and $v_2(D) = pq/4$, so the inequality holds unless $p=q=2$, which is the positive 4-crossing trefoil diagram. □

Theorem 3.4.4 If K is an almost positive knot, then each reduced almost positive diagram D of K has $c(D) \leq 10g(K)+3+2v_2(K)$ crossings.

Proof. D has genus at most $g(K)+1$ by Theorem 2.4.2. Consider the almost positive generator D' from which D is obtained by positive $\bar{t}_2'$-twists. Because almost positive genus 1 diagrams belong to positive knots, $g(D') > 1$, so that by Corollary 3.4.2, D' has at most $10g(D') - 7 \le 10g(K) + 3$ crossings. Moreover, because $g(D') > 1$, D' is not an unknotted twist knot diagram. Thus, $v_2(D') > 0$ by theorem 4.1 of [St3].

Now we apply in a proper order the $\bar{t}_2'$-twists taking D' to D, keeping track of v_2. We know from the Polyak–Viro formula for v_2 that a $\bar{t}_2'$-move augments v_2 except if it is at a crossing p, linked with only two other crossings, q and r, one of which, say q, is the negative one. (In this situation, v_2 is preserved.) Call such a crossing p *thin*. In the proof of Lemma 5.3 of [St3], we observed that q and r then form a trivial clasp in D'. To avoid the elimination of q in D, this clasp must be parallel, and a twist at r must be applied in the sequence of $\bar{t}_2'$-twists taking D' to D. Now we count how many different crossings occur as r in the above situation (for some p). Because $q \overset{\sim}{*} r$, and $\overset{\sim}{*}$ is an equivalence relation, any two possible r and r' are $\overset{\sim}{*}$-equivalent. Then they intersect the same set of other chords, in particular p. But p intersects at most one other chord different from q. So no two different r exist, although for the unique r there may be different p. However, applying first the $\bar{t}_2'$-twist at r, we eliminate all these crossings p as thin crossings (and there are no other ones). So we ensure that all following twists augment v_2. Because $v_2(D') > 0$, at most $v_2(K) = v_2(D)$ twists can be performed (the initial one included). Thus,

$$c(D) \le c(D') + 2v_2(K) \le 10g(K) + 3 + 2v_2(K)\,. \qquad \square$$

Even though the inequalities in Theorem 3.1.1 are optimal in general, they can be improved under additional conditions. One such condition is related to the signature $\sigma(L)$ of the underlying link L (§2.10), in comparison to the maximal Euler characteristic $\chi(L)$ of an orientable spanning surface of L. It is well known [Mu3] that

$$|\sigma(L)| \le 1 - \chi(L)\,, \tag{3.8}$$

at least when L does not bound disconnected surfaces of small genus (see Proposition 2.10.1 and the remark below it), including for example knots or (non-split) positive links. We can say something about the situation when $\sigma(L) \ll 1 - \chi(L)$.

Lemma 3.4.5 Assume we have a diagram D of a link L with $\chi(D) < 0$ and $c_-(D)$ negative crossings, and set

$$k(D) := \frac{1 - \chi(D) - 2c_-(D) - \sigma(D)}{2} \ge 0\,.$$

Then D has at most $-3\chi(D) - \frac{3}{2}k(D)$ $\sim$-equivalence classes, and its underlying generating diagram has at most

$$-5\chi(D) + n(D) - 2 - \frac{1}{2}k(D)$$

crossings.

Proof. If D is a 4-crossing knot diagram, then $k(D)=0$, so this case follows easily. Otherwise, let $\tilde{D}$ be the special diagram obtained from D by Hirasawa's algorithm. Hereby, when applying a move at a separating Seifert circle s, the new strand is placed so as to create $\text{ind}(s)-1$ (and not $\text{ind}(s)$) new negative (non-nugatory) crossings.

We assume first that $\text{ind}(s)\geq 3$. We have shown in the proof of Theorem 3.1.1 that the number of $\sim$-equivalence classes is augmented by

$$2\,\text{ind}(s)-3\geq\frac{3}{2}\big(\text{ind}(s)-1\big)\,, \tag{3.9}$$

and that of the crossing number of the underlying generating diagram at least by

$$2\,\text{ind}(s)-5\geq\frac{1}{2}\big(\text{ind}(s)-1\big)\,. \tag{3.10}$$

For the positive special alternating diagram $\mathring{D}$ obtained from $\tilde{D}$ by crossing changes, we have from [Mu3] $\sigma(\mathring{D})=1-\chi(\mathring{D})=1-\chi(\tilde{D})$, and σ changes at most by 2 under a crossing change. Therefore, we have that $\tilde{D}$ has at least

$$\frac{1-\chi(D)-\sigma(D)}{2} \tag{3.11}$$

negative crossings. Thus $\tilde{D}$ has at least $k(D)$ negative crossings added by the Hirasawa moves (and not inherited from D). Then apply (3.9), (3.10) and Theorem 3.1.1.

This argument shows the assertion if no Hirasawa moves at index-2 Seifert circles are applied. To deal with $\text{ind}(s)=2$, we consider again the modified Hirasawa move (3.4). Then we refine our argument as follows. Mark in each modified Hirasawa move the negative crossing in the additional clasp (this is crossing a in (3.4)). Then we claim that $\tilde{D}$ has (3.11) many (not only negative but) non-marked negative crossings. To see this, it suffices to show that the switch of all marked (positive) crossings in $\mathring{D}$ preserves σ. Now, switching all marked crossings in $\mathring{D}$ and resolving the trivial clasps gives a diagram $\hat{D}$. Clearly, $\hat{D}$ is special and positive. But it also has the same χ as $\mathring{D}$ because modulo crossing changes it is obtained from D by the ordinary (non-modified for $\text{ind}(s)=2$) Hirasawa moves. Thus, $\sigma(\mathring{D})=\sigma(\hat{D})$ follows from Murasugi's result $\sigma=1-\chi$.

Therefore, (3.4) creates (with proper choice between putting the arc above or below) only one non-marked negative crossing but at least two new $\sim$-equivalence classes, and it augments the generator crossing number at least by one. Then (3.9) respectively (3.10) remains valid with the l.h.s. being the augmentation of $\sim$-equivalence classes and generator crossing number respectively, and the parenthetical term on the r.h.s. being the number of (non-marked) negative crossings which contribute to (3.11). □

Remark 3.4.6 Observe that, for knots, instead of σ we can use the Ozsváth–Szabó signature τ, because it coincides (up to a factor) with σ on (special) alternating knots and enjoys (up to that factor) the property (2.33). However, Livingston in fact deduced the Rudolph–Bennequin inequality (2.6) (in the original form of the estimate, without the term $s_-(D)$ on the right) from τ, and so Lemma 3.4.5 becomes trivial for τ. A similar comment applies on the signature s derived from Khovanov homology (for example see Bar-Natan [BN4]).

Note that the signature improvement of Theorem 3.1.1 can be applied to Theorems 3.4.3 and 3.4.4. However, we see in [St8] that, with further tools introduced, the variety of possible modifications and improvements of such inequalities grows considerably so that we cannot discuss each one in detail. We will thus in particular not elaborate much on the use of τ and s and leave the adaptation to an interested reader.

Chapter 4

Generators of genus 4

The methods developed in Chapter 3 allow us to push the compilation of generators for genus 2 and 3 in [St2] one step further and to settle genus 4. Beside its algorithmical merit, this project had the purpose of verifying (and in fact correcting minor mistakes in) the theoretical thoughts on which our approach bases. Several applications presented below in this monograph rely on this compilation.

Table 4.1 gives the distribution of the prime (knot) generators K of genus 4 by crossing number and number of $\sim$-equivalence classes. For maximal generators, we know from our previous work (Theorem 3.1.1) that the crossing numbers in question lie between 21 and 33, but we excluded 21 in [St2] by a computational argument using the property that the Seifert graph must be (3-valent and) bipartite. For 22 crossings, still there are no maximal generators.

Theorem 4.1 There are 3,414,819 prime knot generators of genus 4. Among them there are 1,480,238 special and 1,934,581 non-special ones.

Note that the effort with every new genus increases dramatically, and new theoretical insight was needed to reduce calculations to a feasible magnitude. In our proof, we will explain the algorithm used now. (Still it seems out of scope to attempt the case $g = 5$, at least due to reasons of storing and maintaining the result.)

Proof. For the maximal generators we use the work in [SV]. We first have to determine the planar 3-connected graphs G having knot markings. They can be obtained by a (graphic version of the) γ-construction in [V] from graphs G'' of genus 3:

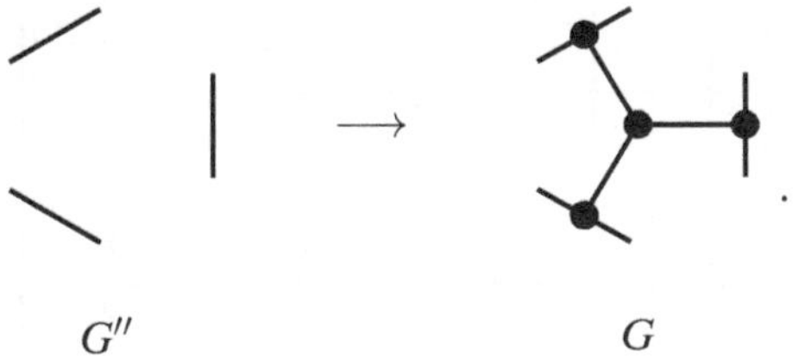

(4.1)

(See §10.3 for more details.)

It is easy to see that if the graph G on the right of (4.1) is 3-connected, then the graph

# ∼ \ c	9	10	11	12	13	14	15	16	17	18	19	20	21	22	23	24	25	26	27	28	29	30	31	32	33	total
8								29																		29
9	1	2	10	28	71	104	147			145																508
10		21	72	210	356	557	660	819	1092			369														4156
11			48	257	766	1791	2942	3832	3080	2804	3188			447												19155
12				55	487	2033	4734	8585	12145	13523	8500	5168	4707			313										60250
13					56	548	3087	9661	19112	27552	31293	27717	14629	5427	3876			111								143069
14						46	590	3519	13251	32388	52870	61747	53398	35540	15787	3173	1827			20						274156
15							41	489	3584	14749	41049	78373	102880	95709	61646	28311	10626	965	465			1				438888
16								27	356	2814	13781	42566	90877	135278	138221	100392	48096	13094	4195	115	49					589861
17									14	231	1854	9704	34955	83859	141210	163710	125842	68515	24978	2942	837					658651
18										4	96	989	5258	20307	56939	110240	150023	136642	75688	27646	8127	172	46			592177
19											4	25	300	2109	8414	25220	57598	92985	105424	77316	29771	5059	1302			405527
20													6	52	401	2181	6905	17039	32977	45891	44939	27879	7828			186098
21															9	36	205	876	2328	4882	8272	10236	9024	5094	1332	42294
total	1	23	130	550	1736	5079	12201	26961	52634	94210	152635	226658	307010	378728	426503	433576	401122	330227	246055	158812	91995	43347	18200	5094	1332	3414819

TABLE 4.1: The number of $\vec{t}_2'$ irreducible prime genus 4 alternating knots tabulated by crossing number c and number of $\sim$-equivalence classes (# $\sim$).

G'' on the left must be at least 2-connected. We determined the graphs G'' from the list of maximal Wicks forms of genus 3 compiled by A. Vdovina. (Note that in these Wicks forms, which are words, an edge is a pair of inverse letters $a^{\pm 1}$ while a vertex is a set of 3 letters $\{a,b,c\}$, such that, for some choice of signs, $a^{\pm 1}b^{\pm 1}$, $b^{\pm 1}c^{\pm 1}$ and $c^{\pm 1}a^{\pm 1}$ are subwords.) From the graphs thus found, we selected only the planar and 2-connected ones and obtained 34 graphs G''.

Then the graphs G on the right of (4.1) are generated by a γ-construction. In applying the move (4.1), some restrictions can be taken into account. For G to become 3-connected, on the left of (4.1) not all 3 segments can belong to the same edge in G'' (although two of them can). Moreover, whenever (a segment of) one in a copy of a multiple (hence double) edge in G'' is affected, the other copy cannot be. Otherwise, the graph G is non-planar or still only 2-connected, or G'' is not 2-connected.

All G obtained were tested for planarity and 3-connectedness (and, of course, isomorphy). We obtained 50 graphs G. This list was later verified by the program of Brinkmann and McKay [BMc] (which I discovered only afterwards and which made it possible to deal with maximal generators of genus 5 and 6; see [St10]).

Then for each graph G on the right of (4.1), we selected all vertex orientations with the orientation of one vertex fixed and bisected edges between equally oriented vertices. The new graph G' corresponds to a knot iff the number of its spanning trees is odd (see [MS]).

All such graphs G' were again tested for isomorphy. We proved in [SV] that the alternating diagrams with these graphs as Seifert graphs do not admit flypes so that there is a one-to-one correspondence between alternating knot, alternating diagram (up to mirroring and orientation) and Seifert graph. Note that isomorphy between the G' is preserved under unbisecting edges so that isomorphic G' must come from the same G, and the isomorphy check for G' can be done for each G separately. After duplications of G' were eliminated, the Dowker notations [DT] were generated, and so 42,294 maximal generators were obtained (which were later confirmed by the calculation in [St10]).

We also checked knot-theoretically that these generators are distinct. The skein polynomial distinguished all except 70 pairs. Among these pairs, the Kauffman polynomial dealt with 62. The remaining 8 pairs were distinguished by the hyperbolic volume.

We know from [SV] that one can obtain the other, not maximal, *special* generators from the maximal ones by resolving reverse clasps. Then for each diagram of a maximal generator we resolved any possible set of (reverse) clasps and identified among all these (alternating) knots the generators of genus 4. They are 1,480,238.

Then we generated all possible special diagrams by flypes. From these diagrams one can then obtain the other (non-special) generating diagrams inductively by the number of separating Seifert circles, using the (reverse to) the moves of Hirasawa (Figure 3.2), and the study of the effect of these moves on generators was carried out in Chapter 3.

We start with a generator diagram D and want to determine all generator diagrams D'

obtained from D by undoing a Hirasawa move. This means that we must identify the segment S in D created by the move. We call S below the *Hirasawa segment*. There are several properties of S that follow from our treatise.

1. This is a segment of odd length (i.e., odd number of crossings on it) and no self-intersection.

2. Let x and y be the first and last crossing on S and call the other crossings *internal*. By the work in Chapter 3, we know that no internal crossings should lie in a non-trivial $\sim$-equivalence class.

3. Moreover, x and y have at most 2 $\sim$-equivalent crossings each. So, prior to undoing the Hirasawa move, we have the option of creating (by a $\bar{t}'_2$-move and possible flypes) two crossings $\sim$-equivalent to x and/or y if x and/or y is in a trivial $\sim$-equivalence class.

4. After the optional application of such $\bar{t}'_2$-moves and flypes, we have the additional condition that the length of the Hirasawa segment S is strictly less than the number of crossings outside S.

5. All possible such segments S were generated, and their crossings removed. Then the resulting Gauss codes of D' were tested for realizability, $\bar{t}'_2$-reducedness, primeness and for having the correct genus (4) and number of separating Seifert circles (one more than in D).

Note that we tested only necessary conditions for S. So some of the diagrams D' we obtain this way may actually not give D by a Hirasawa move. The difficulty is, though, to obtain all diagrams D' we need at least once, and we accomplish this because we reverse all correct Hirasawa moves. The decisive merit of the Hirasawa move is that the way it provides to find all the D' is much more economical in comparison to the previous algorithm in [STV, St2]. Duplicated diagrams D' can be subsequently eliminated by bringing the Dowker notation of any diagram to standard form.

One can then iterate this procedure, undoing (potential) Hirasawa moves (and performing the necessary consistency checks), without the need to flype in between. The process terminates when for some number of separating Seifert circles (8 here, and in general $2g$ by an easy argument) no more (regular) diagrams arise. The flypes are needed after the generation of the diagram lists to identify the underlying knots, and the work is finished. □

To ascertain that this algorithm is practically correct (theoretically it is justified by the proof of Theorem 3.1.1), we verified that the subsets of generators we obtained for ≤ 18 crossings coincide with those that can be selected from the alternating knot tables in [HT] and [FR]. Prior to starting genus 4, we also applied this procedure for genus 3. We checked that the sets of generators we obtained coincide with those generated previously in [St2] by the algorithm described there and in [STV]. While the new implementation dealt with genus 3 about 50 times faster, it was still not before several weeks of labor and computation (partly on distributed machines) that Table 4.1 became complete.

A first small application is easy to obtain and useful to mention. Here and below $O(\dots)$ means "at most ... times a constant independent on n."

Corollary 4.2 There are 680 achiral generators of genus 4. The number of prime achiral alternating genus 4 knots of crossing number n grows like $\frac{43}{2^8\,8!}n^8 + O(n^7)$ for $n \to \infty$ even.

Proof. As in [St2], one must find the achiral generators. The use of Gauss sums (see Proposition 13.3 of [St2]) pre-selects 798 generators, and 680 of them are indeed found achiral by the hyperbolic symmetry program of J. Weeks. The check of the maximal number of $\sim$-equivalence classes of such generators shows it to be 18. This number is attained by 43 of the achiral knots, and none of them has non-trivial $\underset{*}{\sim}$-equivalence classes or admits (even trivial) type A flypes (as explained in Figure 2.6). □

Chapter 5

Unknot diagrams, non-trivial polynomials and achiral knots

5.1 Some preparations and special cases

We now apply the classification of genus 4 generators to prove Theorem 1.5.1. We begin with an easy proof for 2-almost positive diagrams that outlines the later methods.

Proposition 5.1.1 If D is a 2-almost special alternating diagram of the unknot, then D has a trivial clasp.

Proof. We use the Bennequin inequality stated in Theorem 2.4.2. It says that $g(D) \leq 2$.

If D is composite or of genus 1, then the claim is easy to obtain using the genus 1 case [St3]. So assume $g(D) = 2$ and D is prime.

If we have a $\sim$-equivalence class which has crossings of opposite (skein) sign, we either have a trivial clasp, or if not, we have a non-trivial $\sim$-equivalence class which admits flypes. If after the flype and resolution of the clasp we still have a diagram of genus two, we see that we did not start with a diagram of the unknot. By a direct check of the special genus two generators, we verify that this always happens.

We can thus assume that no $\sim$-equivalence class has crossings of opposite (skein) sign. We call a $\sim$-equivalence class positive or negative after the sign of its crossings.

Let $\hat{D}$ be the 2-almost positive (generator) diagram, obtained when removing in D (by undoing $\bar{t}'_2$ twists) crossings in (positive) $\sim$-equivalence classes so that at most 2 crossings in each class remain. Then $\sigma(\hat{D}) \leq 0$.

Assume $\hat{D}$ has two twist equivalent crossings x and y of different sign. A reverse clasp in $\hat{D}$ persists under $\bar{t}_2'$ twists, even up to flypes, so x and y must be $\overset{\sim}{*}$-equivalent. Let us call their equivalence class *bad*.

One can flype the negative crossing $x \in \hat{D}$ to form a parallel clasp with y, and because x is not twisted at in $\hat{D}$ when recovering D, this flype persists in D. Possible $\bar{t}_2'$ twists in D at y create a $(2k+1,-1)$-rational tangle. Then by a $(2k+1,-1) \to (2k,1)$ rational tangle move in D (which is a particular type of wave move),

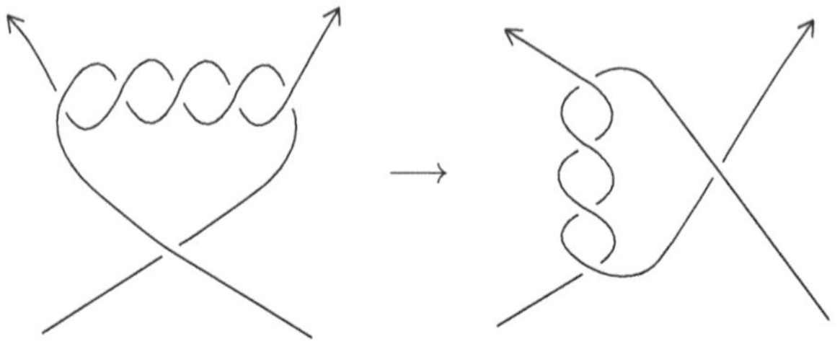

(5.1)

we find a diagram D_1 with one crossing less and $\sigma \le 0$, so that $c(\hat{D}_1) = c(\hat{D}) + 1$, and $\hat{D}_1$ has one bad class less than $\hat{D}$. Still D_1 is ≤ 2-almost positive though not necessarily special. Now repeat such a move until $\hat{D}_1$ has no bad class.

Now by direct check all 2-almost positive genus 2 generator diagrams $\hat{D}$ with $\sigma \le 0$ and no bad class have ≤ 7 crossings and are not special. (Actually $\sigma = 0$, because $\sigma = 4$ on positive genus two diagrams.) They occur for the generators 6_3, 7_6, and 7_7.

Because D_1 is not special, $D_1 \ne D$ and $7 \ge c(\hat{D}_1) > c(\hat{D})$, so $c(\hat{D}) \le 6$. The only special generator coming into question for $\hat{D}$ is 5_1. However, it is easy to check that 2-almost positive 5-pretzel diagrams of the unknot have a clasp, which completes the proof. □

Proposition 5.1.2 If D is a 2-almost positive diagram of the unknot, then D can be made trivial by Reidemeister moves not augmenting the crossing number.

Proof. If D is special, of genus 1, or composite, the claim follows from the almost positive case and the previous proposition. So we assume that D is of genus two, prime and non-special.

We take all non-special genus 2 generators, apply all possible flypes, and switch crossings so that the diagrams are 2-almost positive. We discard all diagrams of $\sigma >$ 0, and from those of $\sigma = 0$ we remove the duplicates.

If we now have crossings of opposite sign forming a (trivial) reverse clasp, this clasp will remain after flypes, and the diagram is simplifiable using a Reidemeister II move. We thus discard diagrams with trivial reverse clasps.

If we have a trivial parallel clasp, we need to have a $\bar{t}_2'$-twist at the positive crossing. We apply this twist in each such clasp and discard all diagrams of $\sigma > 0$. There remain only 3 diagrams, which lie in the series of 6_2. They contain a tangle with a reverse clasp. This tangle can be simplified by Reidemeister III unless there is a twist at the crossings in the reverse clasp. Then we check again that $\sigma > 0$. □

We prepare some arguments that will be useful in the proofs of Theorems 1.5.1, 1.5.2, and 1.5.3.

Lemma 5.1.3 Assume D is a k-almost positive diagram of genus k, and D depicts an achiral knot. Then $P(D) = 1$.

Proof. By Morton's inequality (2.25), we have $\min\deg_l P(D) \geq 0$. So achirality implies that $P(D) \in \mathbb{Z}[l^2]$, i.e., it has only a non-trivial m-coefficient in degree 0. Then [LM, Proposition 21] implies that $P(D) = 1$. □

Lemma 5.1.4 Assume D is a k-almost positive diagram of genus k, and $P(D) = 1$. Then for each pair (a, b) of Seifert circles in D, connected by a negative crossing, there are at least two crossings connecting (a, b).

Moreover, if $k \leq 4$, then for at least one pair (a, b) of Seifert circles connected by negative crossings, there is a positive crossing connecting (a, b).

Using Lemma 2.11.3, we have

Corollary 5.1.5 Any k-almost special alternating diagram with $P = 1$ of genus $k \leq 4$ contains a trivial clasp up to flypes. □

This immediately discards the special genus 4 generators.

Proof of Lemma 5.1.4. We use (2.27). Because Morton's inequality is sharp, we see that $\operatorname{ind}_-(D) = 0$, so that each negative crossing has a Seifert equivalent one. If no negative crossing has a Seifert equivalent positive one, then for $k \leq 4$, the negative crossings lie in at most 2 Seifert equivalence classes. Then an easy observation shows that the diagram D is A-adequate. So by Theorem 2.8.8, $V(D) \neq 1$, and then the same is true for $P(D)$. □

5.2 Reduction of unknot diagrams

We prove here first a slightly weaker version of Theorem 1.5.1, which allows for flypes. We will explain in the next section how to remove the need of flypes.

Proposition 5.2.1 For $k \leq 4$, all k-almost positive unknot diagrams are trivializable by crossing number reducing wave moves and flypes.

Proof. We start with a ≤ 4-almost positive unknot diagram D and will show that one can perform, after possible flypes, a wave move on D. In the following, a wave move is always understood to be one that *strictly reduces* the crossing number $c(D)$ of D. Then we work by induction on $c(D)$. Note that a wave move does not augment either the number of positive or the number of negative crossings (while it reduces one of both). So it does not lead out of the class of ≤ 4-almost positive diagrams. It is possible that the diagram genus becomes larger than 4. However, in this case we know that D cannot be a diagram of the unknot. So we just need to be interested in either showing that D is not an unknot diagram or finding a wave move.

Next, genus and number of negative crossings is additive under connected sum of diagrams. It is then sufficient to work with prime diagrams D with the following additional remark: If we have a composite diagram $D = A\#B$, then flypes in A and B are not affected by taking the connected sum. However, we may have a bridge a in a factor diagram A, which is spoiled when taking the connected sum with B. This can be remedied by displacing B out of a. Such a move can formally be regarded as a flype if in Figure 1.2 one of P or Q has a diagram where one of its strands is an isolated trivial arc. (Let us call this move *factor sliding flype*; it is as in Figure 2.2, except that the tangle is additionally flipped.)

Let D be the initial diagram. Clearly $g(D) \leq 4$. Now let $\hat{D}$ be D with *positive* $\sim$-equivalence classes reduced (by inverse $\bar{t}'_2$ moves) to 1 or 2 crossings. (That is, if a negative $\sim$-equivalence class has > 2 crossings, $\hat{D}$ is not a crossing-switched generator diagram.) Clearly $\sigma(\hat{D}) \leq 0$.

Let k be the number of negative $\sim$-equivalence classes in D (or equivalently $\hat{D}$), and let $\alpha = (a_1, \dots, a_k)$ be the vector of their sizes (number of crossings). We sort the integers a_i non-increasingly, ignoring their order.

Let D' be the positification of a diagram of a generator whose series contains D and $\hat{D}$. That is, D' is obtained from D or $\hat{D}$ by replacing each $\sim$-equivalence class of n crossings with a class of one positive crossing if n is odd and 2 positive crossings if n is even. Then $\sigma(D') \leq 2k$ because twists in a $\sim$-equivalence class do not alter σ by more than ± 2. In particular, we can exclude the case $k = 1$ (i.e., α being one of (4), (3) or (2)) because for a positive diagram D' the property $\sigma = 2$ implies (see end of §2.10) that $g(D') = g(D) = 1$, and then we are easily done by [St4] identifying genus 1 unknot diagrams.

In the first step to generate D, we try to obtain $\hat{D}$ from D' while D' is obtained directly from the generator table by flypes and positification. Flypes on the generator are necessary because they commute with the $\bar{t}'_2$-twists only up to mutations, which we try to avoid (in our stated repertoire of moves simplifying the unknot diagram).

For the possible α, the potential diagrams $\hat{D}$ are determined by crossing switches (and in the case of $\alpha = (3, 1)$, one $\bar{t}'_2$ twist).

It is convenient to check the bound for σ, which is a necessary condition for $\sigma(\hat{D}) \leq 0$, after each crossing switch. Moreover, for given α we select only the D' with $\sigma(D') \leq 2k$. This allows us to discard a lot of possible continuations of the crossing switch procedure, which are understood not to a lead to a proper $\hat{D}$. Also, one should order the $\sim$-equivalence classes of 1 and 2 elements and switch w.r.t. this order to avoid repetitions. Concretely, we proceeded as follows.

- $\alpha = (1,1,1,1)$. Change successively one crossing and check $\sigma \leq 6$, $\sigma \leq 4$, $\sigma \leq 2$, $\sigma \leq 0$.
- $\alpha = (2,1,1)$. Change a reverse clasp and two crossings and check $\sigma \leq 4$, $\sigma \leq 2$, $\sigma \leq 0$.
- $\alpha = (2,2)$. Change two reverse clasps and check $\sigma \leq 2$, $\sigma \leq 0$.
- $\alpha = (3,1)$. Change twice one crossing and check $\sigma \leq 4$ and $\sigma \leq 2$. Then perform a $\bar{t}'_2$ at one of the negative crossings and test $\sigma \leq 0$.
- The cases $\alpha = (2,1)$ and $(1,1,1)$ are analogous.

The diagrams $\hat{D}$ found are relatively few in comparison to the number of generators D'. Also, σ is a mutation invariant. Therefore, to handle the flypes in practice, it is easier to start with one diagram D' per generator, determine whether it gives a $\hat{D}$, and only then to process the other diagrams obtained by flypes from D'. Actually, one can apply the flypes in $\hat{D}$ directly, except for $\alpha = (3,1)$. The latter is easy to handle because the D' with $g(D') = 2$ that we need to consider are few. Thus, let us exclude this in the following.

If $\hat{D}$ has a trivial clasp up to flypes, then all D arising from this $\hat{D}$ will admit a $(2k+1,-1) \to (2k,1)$ rational tangle move (5.1) up to flypes, and so we can discard such $\hat{D}$. Note that only a negative crossing in a trivial $\sim$-equivalence class can form a parallel clasp up to flypes, and because in passing from $\hat{D}$ to D we do not twist at negative crossings, one can flype the negative crossing properly also in D for (5.1) to apply.

Let $\hat{\mathcal{D}}$ be the set of diagrams $\hat{D}$ obtained/maintained so far. In trying to reconstruct D from $\hat{D}$, we distinguish at which positive $\sim$-equivalence classes we apply twists.

Let X be a diagram and $\mathcal{X}$ a set of diagrams. Fix for X the set C_X, consisting of one crossing in each positive $\sim$-equivalence class in X. Let

$$X_* = \left\{ X' : \begin{array}{c} X' \text{ is obtained from } X \text{ by 0 or 1} \\ \text{twist at each crossing in } C_X \end{array} \right\}, \quad \text{and} \quad \mathcal{X}_* = \bigcup_{X \in \mathcal{X}} X_* . \tag{5.2}$$

(The choice to twist or not is made independently for each crossing, so $|X_*| = 2^{|C_X|}$.) The reason we will make use of this construction shortly is the following

Lemma 5.2.2 If a diagram D admits a wave move, and D has a $\sim$-equivalence class of at least 4 crossings, then one can flype and reduce D by a $\bar{t}'_2$-move, so that the resulting diagram D' still admits a wave move. Contrarily, if a diagram D' admits a wave move, and D is obtained from D' by a $\bar{t}'_2$-move at a non-trivial $\sim$-equivalence class, then D' admits a wave move after flypes.

Proof. We focus on the first claim. Let D be presented w.r.t. four of the crossings in the $\sim$-equivalence class T,

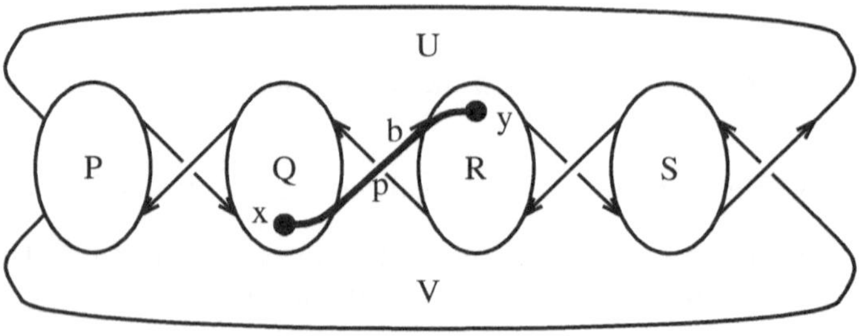

(5.3)

with the (possibly empty) tangles P, Q, R, S being enclosed by these crossings. Let b be a bridge in D that can be shortened by a wave move, and $\operatorname{len} b$ the number of overcrossings b contains. We understand b to start and end on points x and y on the segment (edge) of the diagram line after/before the beginning/ending undercrossing. Now the existence of a wave move means that, if we delete b from the plane curve of D, the distance between the regions X and Y that contain the endpoints x,y of the resulting arc is smaller than $\operatorname{len} b$. Distance between regions X and Y is meant as the shortest length of a sequence of successively neighbored regions, starting with X and ending with Y, counting *one* of X and Y, and neighbored regions means regions sharing an edge.

It is an easy observation that b contains at most one crossing in T (assuming these crossings have all the same sign). Let this be the crossing p between Q and R. (The case that b does not contain any crossing in T is dealt with a similar, though slightly simpler argument.) Now let $\gamma = (X, \dots, Y)$ be a shortest length sequence of neighbored regions connecting X and Y. If γ contains a region W *completely* inside P or S, then it must contain also U and V. However, the distance of U and V is two, and moving from U to V, it is not necessary to pass via W. So γ can be chosen not to contain any regions completely inside P or S (though it may have to contain a region whose corner is one of the crossings in T).

This now means that the flype that makes P and S empty (moving their interior into Q or R) does not change the distance between X and Y. Likewise, it does not affect the bridge b. So b remains too long after such flypes. Now when P and S are empty, one can reduce D by a $\bar{t}'_2$ move, and so we are done with the first statement.

The proof of the second statement is similar. Again a wave move in D' can be chosen so that it does not affect the diagram outside Q, R and p. Then in D one can adjust by flypes the new crossings created by the $\bar{t}'_2$-move at a crossing q to lie outside that portion of the diagram if $p = q$, and entirely within R or Q if $p \neq q$. □

Now from our set $\hat{\mathcal{D}}$ we generate the set $\mathcal{D} = (\hat{\mathcal{D}})_*$. Our understanding is that one can obtain all D from an $X \in \mathcal{D}$ by $\bar{t}'_2$ moves at a non-reduced $\sim$-equivalence class.

The point in this restriction comes from the previous lemma, which says that the applicability of a wave move up to flypes persists under such $\bar{t}'_2$ moves. So if some $X \in \mathcal{D}$ admits a wave move, one can discard all D obtained from X by twisitng at a non-reduced $\sim$-equivalence class. (In fact, the lemma shows that it is necessary to twist only at a crossing in a trivial positive $\sim$-equivalence class, but we will need (5.2) it is given form shortly below.) Similarly, if $\sigma(X) > 0$, the same holds for all D obtained from X by positive $\bar{t}'_2$ twists. Such X can be likewise discarded.

Therefore, discard all diagrams X in $\mathcal{D}$ admitting a wave move or having a positive signature, obtaining a subset $\mathcal{E}$ of $\mathcal{D}$. We deal with $\mathcal{E}$ using the maximal degree of the Jones polynomial. This procedure was described in [St14]. What we need in our case is the following property.

Lemma 5.2.3 (regularization; [St14]) Assume that there is an integer n such that

$$n = c(D'') - \max\deg V(D'') \text{ for all } D'' \in D_*. \tag{5.4}$$

Then (5.4) holds for all diagrams D' obtained by positive $\bar{t}'_2$-twists and flypes from D.

The proof is a simple inductive application of the skein relation of V. We called in [St14] property (5.4) *regularity*, and its verification *regularization*. We will use below the same terms. (There is a similar property for P, which we will encounter in some form in Chapter 7.)

We check first that $\max\deg V(E) > 0$ for all diagrams $E \in \mathcal{E}$. Then split $\mathcal{E}$ into sets $\mathcal{E}_n$ of diagrams E, according to $n = c(E) - \max\deg V(E)$. (In practice, the occurring values of n are between 3 and 6.) Now D is supposed to be obtained from some diagram in some $\mathcal{E}_n$ by some number of positive $\bar{t}'_2$-twists. To rule this out, build $(\mathcal{E}_n)_*$ and check for each $E \in (\mathcal{E}_n)_*$ that $n = c(E) - \max\deg V(E)$. Using the skein relation of V one can then show that this property is preserved under further positive $\bar{t}'_2$-twists. So if D is a diagram not admitting a wave move, then

$$\max\deg V(D) = c(D) - n \geq c(E) - n = \max\deg V(E) > 0\,.$$

Thus, no unknot diagram can occur as D. With this Proposition 5.2.1 is established. □

5.3 Simplifications

There are a few shortcuts in the above procedure that we mention. In building $\mathcal{E}$ from $\mathcal{D}$, one can also discard all semiadequate diagrams X. Semiadequacy persists under twisting, and by [Th] (or by Theorem 2.8.8) the unknot has no non-trivial semiadequate diagrams. Because we will come to talk about it in the sequel, we call this step shortly the *semiadequacy shortcut*. For knots with $P = 1$, it can always be applied.

When switching crossings one by one to obtain $\hat{D}$ from D', for $c_-(D) = g(D) \in \{3,4\}$, we can use Lemma 5.1.4. We discard special generators immediately and for non-special generators switch only crossings that have a Seifert equivalent one. For $g = 4$, we are left, after flyping and discarding $\hat{D}$ with trivial clasps, with 31 possible diagrams. Using the semiadequacy shortcut, we find $\#\mathcal{E} = 4$, and regularization applies with $n = 5$.

Another shortcut comes from the Rudolph–Bennequin inequality (2.6). It implies the following:

Lemma 5.3.1 If D is a diagram of a knot K, and k the number of $\sim$-equivalence classes of D which contain a negative crossing, then $g_s(K) \geq g(D) - k$.

Proof. We flype D so that between the n negative crossings in each such $\sim$-equivalence class, we have $n-1$ negative Seifert circles (of valence 2). The estimate then follows from (2.6). □

This means that for given α (index of distribution of negative crossings into $\sim$-equivalence classes) of length k, one needs to consider only D' with $g(D') \leq k$. We call this step the *Rudolph–Bennequin shortcut*.

The reason for not including these shortcuts into the proof of Proposition 5.2.1 is that in the modifications of the proof for the related theorems their legitimacy will vary from case to case.

We now set out to prove Theorem 1.5.1 in its full form. The basic point of removing flypes from Proposition 5.2.1 lies in the construction (5.2) and Lemma 5.2.2. We first improve Lemma 5.2.2.

Lemma 5.3.2 Assume a diagram D is given in the presentation (5.3), with at least 2 tangles R, Q. Assume D admits a reducing wave move shrinking a bridge $b \subset R \cup Q \cup \{p\}$ to a bridge b'. Then this move can be chosen so that b lies entirely within R or Q, or in $R \cup \{p\}$ or $Q \cup \{p\}$.

Proof. If b does not contain p, then b lies entirely within R or Q. So assume again that b passes p. It is obvious that b' can be chosen not to pass any crossing outside of $R \cup Q \cup \{p\}$. If b' passes p, then the wave move between b and b' would split into two parts, one which modifies the part of b between x and p and another one which acts on the remaining part between p and y. At least one of these would be a reducing move, and so we are done.

Thus, we may assume that b' does not pass p. Then there are 2 options how b' looks outside R and Q:

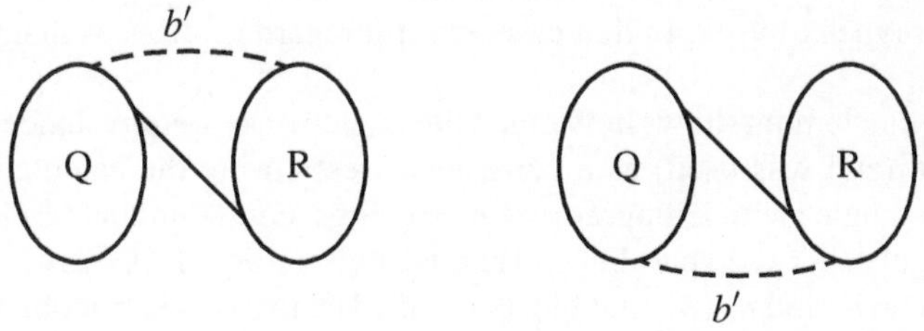

They are symmetric, so look just at the first one. By drawing a circle γ like

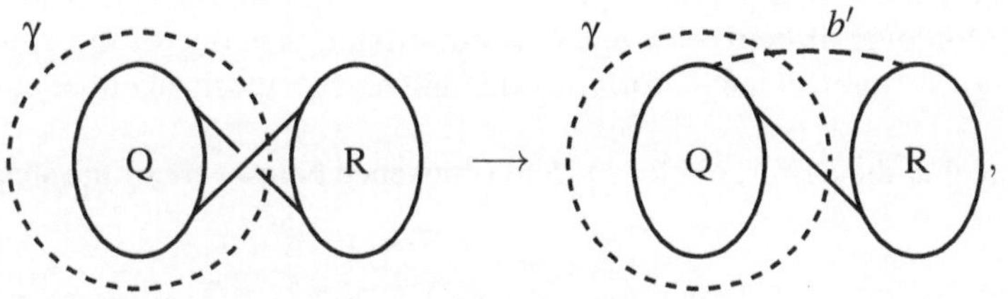

we see that again γ splits the wave move into two parts, for which the bridge lies within $Q \cup \{p\}$ and R respectively, and at least one of which is reducing. □

Now we need a more delicate treatment of the freedom to apply flypes at a crossing of a $\sim$-equivalence class. It will be convenient to slightly change the nomenclature of tangles from (5.3).

Definition 5.3.3 Let the (flyping) degree $\deg C$ of a $\sim$-equivalence class C in D be the maximal number of tangles T_i in a presentation of D like (5.5) (where it is 4), such that no T_i is trivial (all contain at least one crossing), no T_i contains crossings in C, and each crossing not contained in any tangle T_i belongs to C. We call such T_i *essential* (w.r.t. C).

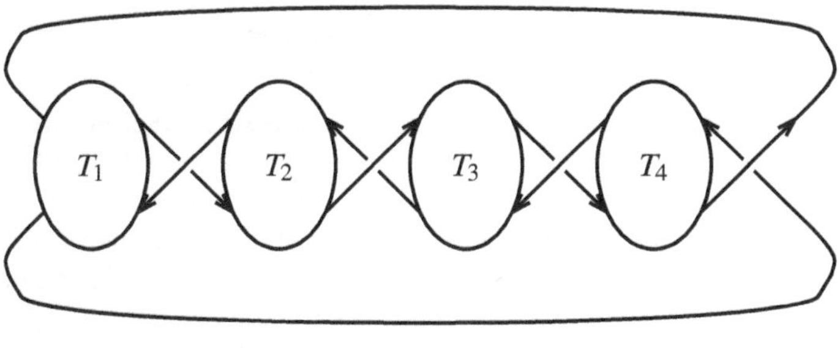

(5.5)

Hereby, we insist on the strand orientations given in (5.5). We call C *(flyping) inactive* if it has degree 1. (If $\deg C = 0$, then D is a reverse $(2,n)$-torus link diagram, which can be easily dealt away with.) Note that in a trivial $\sim$-equivalence class, a crossing may admit type A flypes. In that case, we still regard its class as inactive.

The following lemma shows how much the degree of a $\sim$-equivalence class can grow. It is useful (and was used) as a consistence test during the calculation. An *n-twist tangle* is a tangle (with a diagram) of n crossings, containing $n-1$ clasps. Again we can distinguish between parallel and reverse twists if $n>1$. A *(n,m)-pretzel tangle* is the sum of an n- and m-twist tangle, performed so that we do not obtain a $n+m$-twist tangle.

Lemma 5.3.4 In a knot diagram of genus g, each $\sim$-equivalence class C has degree at most g. Moreover, if $\deg C = g$ and C is non-trivial, then one of the T_i is a twist tangle, or a (n,m)-pretzel tangle for n,m odd (and reverse twists in either case).

Proof. Note that the Seifert circles in (5.5) (thickened below) are of the shape

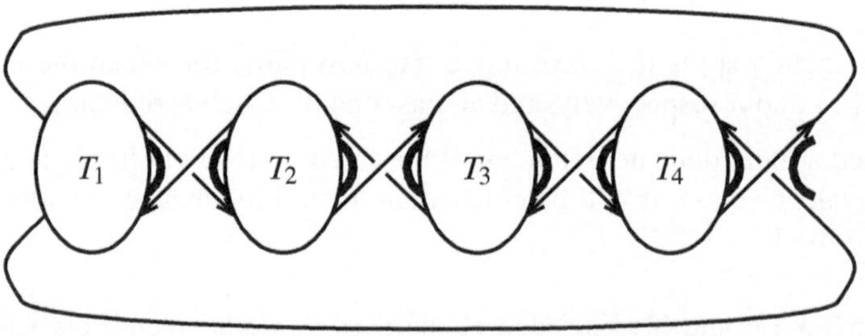

We look at the Euler characteristic. First $\chi(D) = 1-2g$. We apply next the replacement

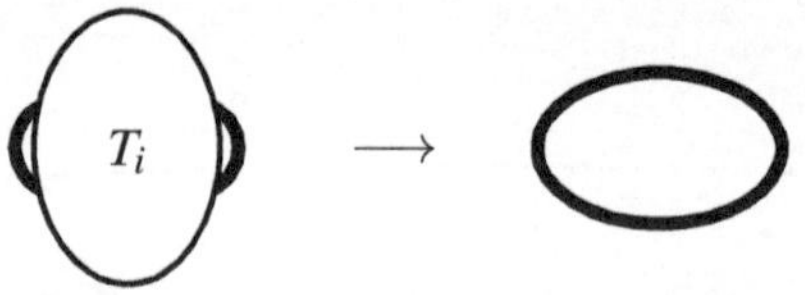

(5.6)

(in other words, "emptying out" T_i and making it into a single Seifert circle). One observes that, because T_i is essential, (5.6) augments χ by at least 1, and by exactly 1 only if T_i is a (n,m)-pretzel tangle or a $2n$-twist tangle. Connectivity of D implies that in former case n,m are odd, and in either case that at most one such T_i can occur. Thus (5.6) for all but at most one i augments χ by at least 2. Finally, after all instances of (5.6) are performed, the resulting diagram D_0 is still connected, so $\chi(D_0) \le 1$.

The first claim of the lemma then follows by a simple estimate. The second claim follows similarly because if C is non-trivial, then $\chi(D_0) = 0$. Thus, at least once in (5.6), χ must go up only by 1. □

Definition 5.3.5 Let X be a diagram. Fix for X a set C_X, consisting of one positive crossing in each non-negative $\sim$-equivalence class in X. (Non-negative means here that the class contains at least one positive crossing.) Let $p : C_X \to \mathbb{N}$ be a function, defined on $c \in C_X$ depending on its $\sim$-equivalence class C as follows:

$$p(c) = \left\{ \begin{array}{cl} 1 & \text{if } C \text{ is trivial and inactive,} \\ 0 & \text{if } C \text{ is non-trivial and inactive,} \\ 0 & \text{if } C \text{ is active and } |C| \geq \deg C, \\ \left\lfloor \dfrac{\deg C - |C| + 1}{2} \right\rfloor & \text{if } C \text{ is active and } |C| < \deg C. \end{array} \right\}.$$

Then let

$$X_+ = \left\{ X' : \begin{array}{c} X' \text{ is obtained from } X \text{ by } 0, \ldots, p(c) \\ \text{twists at } c \text{ for each } c \in C_X \textit{ and type} \\ \textit{B flypes}, \text{ so that } X' \text{ has no trivial clasp} \end{array} \right\}.$$

(The choice of twists is again understood to be independent for each crossing, so that there are $\prod_c (p(c)+1)$ choices. Now, however, $|X_+|$ will in general be larger, due to the flypes.) Again if $\mathcal{X}$ is a set of diagrams, we let $\mathcal{X}_+ = \bigcup_{X \in \mathcal{X}} X_+$.

The next lemma shows up to how many crossings we must twist in each $\sim$-equivalence class to be sure that reducing wave moves will persist (without need of flypes) if more crossings are added.

Lemma 5.3.6 Assume that for a generator D, every diagram in D_+ admits a reducing wave move. Then so does every diagram in its series $\langle D \rangle$ obtained by type B flypes and positive $\bar{t}'_2$-twists from D.

Proof. Let $D' \in \langle D \rangle$. We fix a $\sim$-equivalence class C in D' that contains at least one positive crossing. Then we will obtain below a diagram $D'' \in \langle D \rangle$ from D' by flypes and removing pairs of crossings in C by undoing twists. In this diagram D'' the class C will have at most $\deg C + 1$ crossings if C is active, and at most 3 crossings if C is inactive. We will argue that if D'' admits a reducing wave move, then so does D'. The lemma follows then by induction on the $\sim$-equivalence classes of D.

To find this diagram D'', first assume that C is inactive. Let $T = T_1$ be the complementary tangle to C. We may then assume that C is positive because otherwise we have a trivial clasp. (In the absence of flypes, one cannot put tangles "between" clasps.) If $|C| \leq 3$, we may set $D'' = D'$, so let $|C| \geq 4$. Now remove pairs of crossings in C by undoing twists until 2 or 3 crossings remain. Let this be D''.

If D'' admits a reducing wave move, then Lemma 5.3.2, applied on $T_1 = T$ and T_2 consisting of the $|C| - 2$ internal crossings in C, shows that this move avoids T_2. (Note that, unlike Definition 5.3.3, Lemma 5.3.2 makes no assumption whatsoever on T_i.) Then T_2 can be replaced by arbitrary many crossings in C, and our claim for D'' is justified.

Next let C be active. We assume that $|C| \geq \deg C + 2$, and that between the T_i only crossings of the same sign occur in D'. (Otherwise we have again a trivial clasp.) We call such a collection of crossings a *group*. We attach a non-zero integer to a group by saying that it is a group of $-t_i$ if it has t_i negative crossings, and it is a group of t_i if it has t_i positive crossings.

If there is a triple of crossings, i.e., a group of t_i for $|t_i| \geq 3$, then we can undo a $\bar{t}'_2$-twist. Each T_i will have a crossing of the same sign on either side, so by Lemma 5.3.2 a wave move in the simplified diagram would give a wave move in the original diagram.

If there are two pairs of crossings, then we remove one crossing in each pair and flip the portion of the T_i on one side around. (This is a flype followed either by undoing a twist, or removing a trivial clasp.) The same argument applies, only that bridges might have turned into tunnels. (Keep in mind that we regard tunnels as equivalent to bridges.)

With this we managed to simplify C by two crossings, so to find D'', repeat this until $|C| - \deg C \in \{0, 1\}$.

Let us finally briefly argue why we can assume that D'' has no trivial clasps. First, we assumed that in D' every group of a $\sim$-equivalence class is signed and constructed D'' so that it has the same property. This is why D'' has no trivial reverse clasp. If D'' had a trivial parallel clasp, then the two crossings in it would be in a trivial and inactive $\sim$-equivalence class. Thus, the only difference between D'' and D' in that classes could be that at the positive crossing, some $\bar{t}'_2$-twists are applied. We would then have a variant of the move (5.1) in D'. □

The following modification of the previous lemma is the analogue of the test given in the paragraph after the proof of Lemma 5.2.2, which we apply now to avoid flypes.

Lemma 5.3.7 Assume that D is a (not necessarily alternating) generator diagram or obtained from such a diagram by one negative $\bar{t}'_2$-twist and possible flypes. Assume that each diagram D'' in D_+ admits a reducing wave move, or $\sigma(D'') > 0$. Then some of these two alternatives applies to every diagram $D' \in \langle D \rangle$ obtained by type B flypes and positive $\bar{t}'_2$-twists from D.

Proof. Remark that in the previous proof undoing a $\bar{t}'_2$-twist at a positive crossing, as well as a flype, does not augment σ.

Then we only consider the situation where we need to undo a negative $\bar{t}'_2$-twist. Let C again be the $\sim$-equivalence class modified in passing from D' to D''. Following the argument in the previous proof, we see that a negative $\bar{t}'_2$-twist could occur (after a possible flype) only if in C we have a group of $t_i \leq -3$, or two groups of $t_i = -2$. Then we have ≥ 3 negative crossings (at at least one positive crossing) in C.

So assume that D has a $\sim$-equivalence of ≥ 3 negative crossings. This means, by the Rudolph–Bennequin shortcut, that we need to consider only generators of genus at most 2. The description of unknot diagrams of genus 1 is easy.

We thus assume $g(D') = 2$. By Lemma 5.3.4, we have $\deg C \leq 2$. We assumed that C has a positive crossing, and thus if D' has no trivial clasp, at least one of the two groups in C has $t_1 > 0$. Because of the negative crossings in C, the other one must have $t_2 \leq -3$, and in particular $\deg C = g(= 2)$. The second part of the lemma then shows that one of the T_i is a twist tangle respectively pretzel tangle. (One can alternatively use Theorem 2.11.11 directly.) This tangle, say T_1, is connected by a crossing in C on one side in a non-alternating way. Thus, for a twist tangle T_1, we have a version of the move (5.1) (in which the twist tangle would have an even number of crossings and the orientation would be slightly different).

For a pretzel tangle T_1, we can apply a similar move if one of m or n is equal to 1. If $m,n \geq 3$, we will argue that the diagram is knotted. We have in D' a tangle like

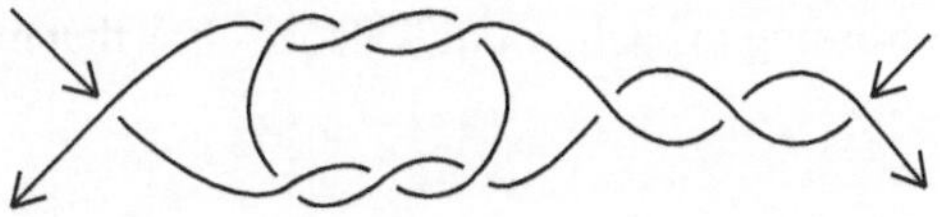

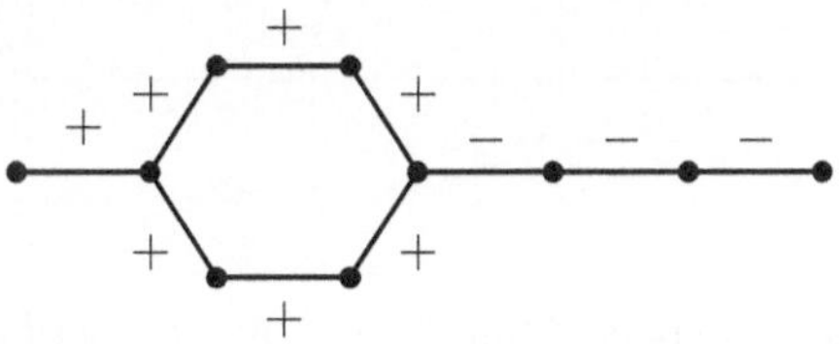

Its Seifert graph is shown below. We now use the inequality (2.27) of Murasugi–Przytycki. (The problematic cases discussed below in §7.2 do not occur here.) By successively contracting the leftmost vertex of the negative edges, we see that $\text{ind}_-(D') \geq 3$. Then (2.27), slightly rewritten using §2.1, yields

$$\min\deg_l P(D') \geq 2g(D') - 2c_-(D') + 2\,\text{ind}_-(D') \geq 2\,.$$

The skein polynomial thus shows that the unknot cannot occur in such a diagram. □

Proof of Theorem 1.5.1. The case $k = 1$ was known, and $k = 2$ can be easily concluded from the proof of Proposition 5.1.2. Let us thus assume $k = 3,4$. Moreover, by Theorem 2.11.10, we may assume $g(D) \geq 2$.

The work we did allows us to now proceed similarly to Proposition 5.2.1. We consider generators of genus 2 to k. The factor slide allows us to discard composite generators and Corollary 5.1.5 special ones of genus k. We take the positification of one diagram per generator. We switch crossings (successively testing σ to save work) to find all k-almost positive diagrams of $\sigma \leq 0$. We must here, however, allow for crossings of opposite sign in the same $\sim$-equivalence class, including for trivial clasps.

Our understanding is that each diagram we have to treat is obtained from the generator diagram first by type A flypes, then $\bar{t}'_2$-twists, and finally type B flypes. Thus, first we apply type A flypes.

Additionally, we must consider for $k = 4$ the diagrams, where the negative crossings split into $\sim$-equivalence classes $(3,1)$. We take all genus 2 generators, apply type A flypes, make positive, switch two crossings, and test $\sigma \le 2$. Then we apply one negative $\bar{t}_2'$-twist, and test $\sigma \le 0$.

Let $\mathcal{D}$ be the set of diagrams that we obtained so far. Our understanding is now that every diagram we need to consider is obtained from those in $\mathcal{D}$ by positive $\bar{t}_2'$-twists and type B flypes. For each diagram D in $\mathcal{D}$, we then replace D_* in the context of (5.2) by D_+. (This then removes the diagrams with trivial clasps.) We discard all diagrams in $\mathcal{D}_+$ admitting a reducing wave move or having $\sigma > 0$.

Finally, we apply Jones polynomial regularization (Lemma 5.2.3) on the remaining set $\mathcal{E}$. Here still $(\mathcal{E}_n)_*$ must be used (an *not* $(\mathcal{E}_n)_+$). However, now crossings of either sign may occur in a $\sim$-equivalence class, so the set C_X in (5.2) must be understood to contain one positive crossing in each $\sim$-equivalence class that has a positive crossing.
□

5.4 Examples

Example 5.4.1 It is to be expected that the simplification of unknot diagrams grows considerably more complicated when the number of negative crossings increases. We give a few examples illustrating this.

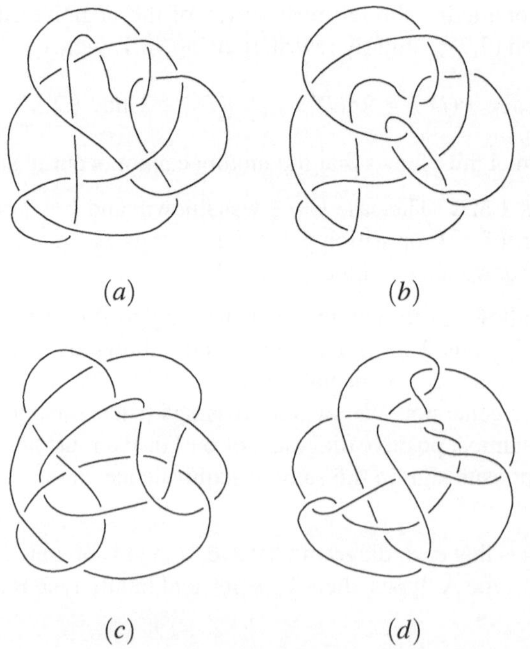

(5.7)

The 5-almost positive 14 crossing diagram (a) does not allow a reducing wave move, and so Theorem 1.5.1 fails in this form for $k = 5$. Our example allows a reducing wave move after a flype. We then checked that up to 18 crossings, all 5-almost positive unknot diagrams were still found to admit a reducing wave move after flypes, so that at least Proposition 5.2.1 might be true for $k = 5$. For 6-almost positive diagrams, even flypes are not enough. The 15 crossing diagram (b) admits a flype, but no wave move applies in any of the two diagrams obtained by flypes. The diagram (c) is similar, though now no flypes apply either. The 16 crossing diagram (d) is like (c) but is additionally special.

From the point of view of alternation, we have from Theorem 1.5.1:

Corollary 5.4.2 For $k \leq 4$, a special k-almost alternating unknot diagram is trivializable by crossing number reducing wave moves and factor slides. □

Example 5.4.3 We made again some experiments to gain evidence as to how (in)essential the specialty assumption is. Tsukamoto's Theorem 2.6.3 for $k = 1$ indicates that flypes must be included. We found that all 2-almost alternating unknot diagrams up to 18 crossings admit, after possible flypes, a reducing wave move, and in fact such a move that leads *to a 2-almost alternating diagram.* In contrast, there is a 3-almost alternating 17 crossing diagram (on the right) which does not admit either a flype or a (whatsoever) reducing wave move.

As already highlighted, both Corollary 5.4.2 and the above computation deserve to be accompanied by the warning that wave moves do not (in general) preserve the property $\leq k$-almost alternating. (In particular, we did not claim in the corollary that diagrams obtained intermediately after moves during the process of reduction are either special or ≤ 4-almost alternating.) Thus, there seems some meaning in asking what moves in $\leq k$-almost alternating diagrams should naturally play the role of wave moves in $\leq k$-almost positive diagrams.

A common feature of these examples is that one can still simplify them by (reducing) wave moves after first performing some crossing number preserving ones. There seems some folklore belief that this might be true in general.

Conjecture 5.4.4 Crossing number preserving and crossing number reducing wave moves in combination suffice to trivialize unknot diagrams.

I only had an imprecise source for this conjecture, which also reports that it was checked up to relatively large crossing number (above 20). Within my computing capacity, I was able to confirm it up to 17 crossings. We explained already that (because factor sliding is a preserving wave move), Theorem 1.5.1 in particular settles the conjecture for ≤ 4-almost positive diagrams.

Remark 5.4.5 We do not know if factor sliding is essential in Theorem 1.5.1. The calculation we did was based on the Dowker notation [DT]. It would make a treatment of composite diagrams indefinitely more complicated, too much for the minor (potential) benefit of removing the factor sliding.

5.5 Non-triviality of skein and Jones polynomial

Proof of Theorem 1.5.2. The non-triviality result for the skein polynomial is almost immediate from the proof of Proposition 5.2.1. Assume $P = 1$. Then in particular $V = 1$, so K is not semiadequate by Theorem 2.8.8, and $\Delta = 1$, so $\sigma = 0$. Also by $\min\deg_l P = 0$ and Morton's inequalities a ≤ 4-almost positive diagram D must have $g(D) \leq 4$. The rest of the argument and calculation is as in the proof of Proposition 5.2.1. Here the semiadequacy shortcut is allowed but not the Rudolph–Bennequin shortcut. This deals with the statement in Theorem 1.5.2 about the skein polynomial.

We now make more effort for the more interesting problem regarding V. The cases $k \leq 1$ are well known. First assume $k = 2$. If $V = 1$, then by [St6] a 2-almost positive diagram D has $g(D) \leq 3$. Also $V = 1$ implies $\det(L) = 1$, so that from (2.37) we have $8 \mid \sigma(L)$, and using $g(L) \leq g(D) \leq 3$ and (3.8), we have again $\sigma = 0$. (We will invoke (2.37) in this way several times below.)

If D is composite, then one is easily done using the result for almost positive knots in [St6]. (Note that when $\max\deg V > 0$, then V is not a unit in $\mathbb{Z}[t^{\pm 1}]$; see [J2, §12].) If D has genus 1, then K is a rational or pretzel knot, so it is a Montesinos knot. By [LT], it is semiadequate, and thus by Theorem 2.8.8, $V \neq 1$.

So start with positive generators D' of genus 2 or 3 and $\sigma \leq 4$, switch two crossings to find $\hat{D}$, and check $\sigma(\hat{D}) \leq 0$ and that $\hat{D}$ is not semiadequate. We again obtain a set $\hat{\mathcal{D}}$. We again construct $\mathcal{D} = (\hat{\mathcal{D}})_*$ and discard semiadequate, positive signature or wave move admitting diagrams X therein. Then we again obtain $\mathcal{E}$ and apply a $\max\deg V$ regularization.

For the rest of the proof we consider $k = 3$. By [St6] we have $g(D) \leq 5$. Again the composite case can be easily reduced to the prime one, so assume D is prime.

Case 1. Clearly we must first dispose of the (computationally impossible) case of $g(D) = 5$.

If the 3 negative crossings are divided into Seifert-equivalence classes 1-1-1, then the situation in part (a) of Figure 5.1 occurs 3 times; possibly the loops on the right of (a) are connected to one. (Here we must take into account that the Seifert graph of D is bipartite, so any cycle of non-parallel edges has a length of at least 4.) So the A-state of D has $|A(D)| = s(D) - 3$ loops. Converting this to V using (2.20), we see that it contributes in degree $m(D) = g(D) - 3 = 2$. So $\min\deg V \geq 2$, and $V \neq 1$.

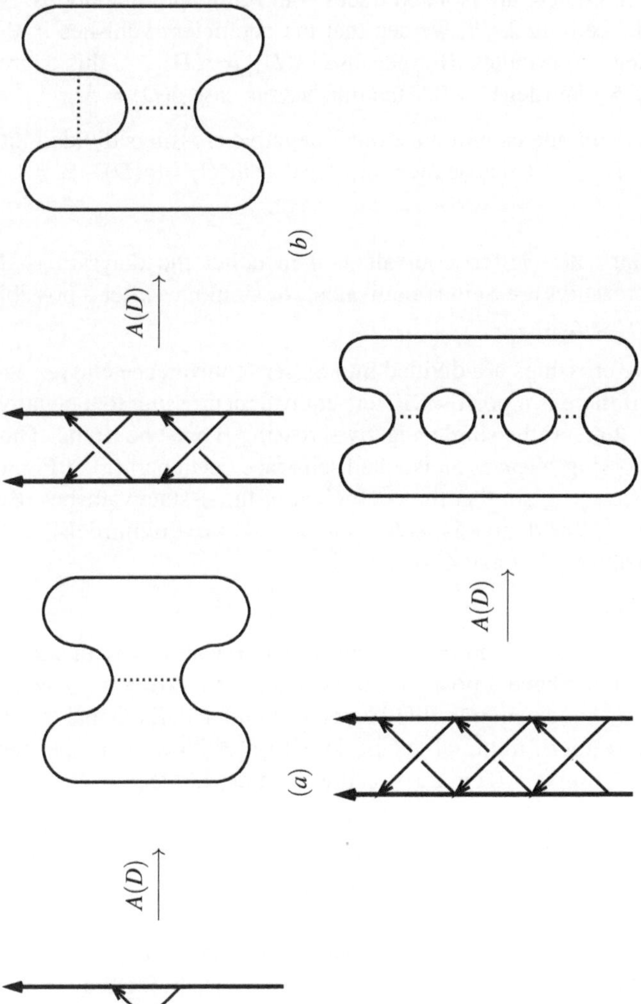

FIGURE 5.1: Some possible loops in the A-state of D. The left sides display the location of crossings for a pair of Seifert circles (thickened), and the right part displays how the loops are connected in $A(D)$.

If the 3 negative crossings are divided 2-1 in Seifert-equivalence classes, the same argument and $|A(D)| = s(D) - 1$ shows that we have $\min\deg V \geq m(D) = 1$.

Then all 3 negative crossings are in a common Seifert-equivalence class. So we have a loop in $A(D)$ like in part (c) of Figure 5.1. Now if there are no more (positive) crossings in that Seifert-equivalence class, D is A-adequate. If there are such crossings, then in the A-state of D there are isolated traces connecting the same loops. By the analysis of [BMo] in Lemma 2.8.9, we see that the coefficient vanishes in the degree in which the A-state contributes. But because $|A(D)| = s(D) + 1$, this degree is $m(D) = g(D) - 5 = 0$. So $\min\deg V > 0$. This finishes the case $g(D) = 5$.

Case 2. Let $g(D) = 4$. Again one cannot have the 3 negative crossings divided into Seifert-equivalence classes 1-1-1 because then $\min\deg V \geq m(D) = g(D) - 3 = 1$.

Case 2.1. D is special.

If all 3 negative crossings are Seifert-equivalent, then either the diagram is A-adequate (if no further crossings are Seifert-equivalent) or simplifies after a possible flype (if there are such crossings).

So assume the 3 negative crossings are divided into Seifert-equivalence classes 2-1. Now D would simplify if there is a positive Seifert-equivalent crossing to a negative one. Otherwise the class that has the single negative crossing p must be trivial. Then in the A-state of D the crossing p leaves an isolated self-trace (as in part (a) of Figure 5.1). By Lemma 2.8.9, we see again that the coefficient of the A-state vanishes. But because $|A(D)| = s(D) - 1$, this degree is $m(D) = g(D) - 4 = 0$. So $\min\deg V > 0$. This finishes the case that $g(D) = 4$ and D is special.

Case 2.2. D is non-special.

Case 2.2.1. The 3 negative crossings are Seifert-equivalent. Because we can assume D is not A-adequate, we must have a positive Seifert-equivalent crossing. Now the canonical surface of D then has a compressible Murasugi summand. By Gabai's work [Ga2, Ga3], this surface is not of minimal genus, so $g(L) \leq 3$. Now again, because $8 \mid \sigma$ from (2.37), the condition $\sigma = 0$ emerges, and we can use a σ test.

We need to start only with positivized generator diagrams D' for which $\sigma(D') \leq 6$. (This drops drastically the work—see the proof of Proposition 6.1.) To obtain $\hat{D}$, we switch 3 crossings in a Seifert-equivalence class of at least 4, and check if $\sigma(\hat{D}) \leq 0$. (Because σ is unaltered under mutation, one diagram per generator knot will do.) 10 diagrams $\hat{D}$ remain; after discarding semiadequate diagrams 4 remain. Then we build $\mathcal{X}_*$ as in (5.2) for the set $\mathcal{X}$ of these 4 diagrams. After discarding diagrams in $\mathcal{X}_*$ admitting a wave move and such of $\sigma > 0$, the diagram set becomes empty.

Case 2.2.2. The 3 negative crossings are in Seifert-equivalence classes 2-1. The degree $m(D)$ in (2.20) of the A state is again 0, and by Lemma 2.8.9, we see that, for $\min\deg V = 0$, there must be positive crossings Seifert-equivalent to the single negative crossing in the one class (as in part (b) of Figure 5.1) and no positive crossings Seifert-equivalent to the pair of negative crossings in the other class.

Again Gabai and $8 \mid \sigma$ imply that $\sigma = 0$. So we need only the positivized generators with $\sigma(D') \leq 6$. It turned out that after switch of the pair of crossings in a Seifert-equivalence class of 2, we never had $\sigma \leq 2$. In particular, we cannot build a $\hat{D}$ by one more crossing switch so that $\sigma(\hat{D}) \leq 0$. So we are done.

Case 3. $g(D) \leq 3$. Now $8 \mid \sigma$ again implies $\sigma = 0$. So we could proceed as for Theorem 1.5.1, with the small difference that we avoid the use of Lemma 5.1.4. □

Remark 5.5.1 For $k = 2$ and the Alexander polynomial, a similar result follows from [St7] because we showed that 2-almost positive knots of zero signature are only the twist knots (of even crossing number with a negative clasp). Later we will derive this result in a far more elegant way from our setting, using some of the work in [St12]; see Chapter 6.

Example 5.5.2 The $(-3,5,7)$-pretzel knot is an example of a knot with trivial Alexander polynomial, which is (by the preceding remark) 3-almost positive. We do not know whether it is the only one. (For this particular case one can, again, use the methods of this paper to easily verify 3-almost positivity directly rather than appealing to [St7].)

Corollary 5.5.3 If L is a ≤ 4-almost positive smoothly slice knot, then $V(L) \neq 1$.

Proof. Let D be a ≤ 4-almost positive diagram of K. Theorem 1.5.2 easily deals with the case that D is composite (because, as noted it its proof, if $V \neq 1$, then V is not a unit). Considering D to be prime, we have $\sigma = 0$, and Lemma 5.3.1 implies that $g(D) \leq 4$. Then a calculation like in the proof of Theorem 1.5.2 applies. □

In special diagrams, one can extend Theorem 1.5.1 to $k = 5$, if one allows for flypes, and similarly the part of Theorem 1.5.2 for the skein polynomial.

Theorem 5.5.4 Let D be a k-almost special alternating knot diagram for $k \leq 5$. Then the following conditions are equivalent:

(1) D is unknotted,

(2) $P(D) = 1$,

(3) D can be trivialized by reducing wave moves and flypes.

Remembering the diagram (d) in (5.7), we see that the theorem is not true for $k = 6$. On the other hand, for $k = 5$ it might we true for k-almost positive diagrams (without the assumption the diagram to be special) as explained in Example 5.4.1. The warning formulated after Example 5.4.3 is valid here, too.

Proof. Clearly we need to prove only (2) $\Rightarrow$ (3). We may further assume that $k = 5$ because the cases $k \leq 4$ are contained in Theorems 1.5.1 and 1.5.2. Next we may assume that D is prime because (again, for example using (2.10) and properties of V), if $P(D)$ is a unit in $\mathbb{Z}[l^{\pm 1}, m^{\pm 1}]$, then $P(D) = 1$.

Because $k = c_-(D) = 5$, we have $g(D) \leq 5$ by (2.25). Next, $P = 1$ implies $\Delta = 1$, which in turn implies $\sigma = 0$, and the signature test is allowed. Likewise is the semiadequacy test because $P = 1$ implies $V = 1$.

Assume $g(D) = 5$. Because D is special, by Lemma 2.11.3, each $\overset{\sim}{*}$-equivalence class is signed, or we can simplify D after a flype. By (2.27), because $\min\deg_l P(D) = 1$ and $k = c_-(D) = 5$, we must have $\mathrm{ind}_-(D) = 0$. This means that there is no negative $\overset{\sim}{*}$-equivalence class of a single crossing. Thus, the 5 negative crossings in D must lie either in a single negative class or split into two such classes 2-3. In either case, D is A-adequate, and so $V(D) \neq 1$ by Theorem 2.8.8.

Then we have $g(D) \leq 4$, and can use the generator table. If $g(D) = 1$, then by Theorem 2.11.10, the claim can be checked easily. (All such D are semiadequate, except if they have a trivial clasp.)

We consider thus $g(D) = 2, 3, 4$. The rest of the proof is as for Proposition 5.2.1, though with a fifth negative crossing. It may be necessary (and thus we do) consider non-special diagrams. Again, at most one negative $\bar{t}'_2$-twist must we dealt with. In the case of such a twist, we must here test all the generators of genus 2 and 3 and those of genus 4 whose positification has $\sigma \leq 6$. These cases are easily finished.

We consider the case of no negative $\bar{t}'_2$-twist. We proceed as before until we build again $\mathcal{D} = (\hat{\mathcal{D}})_*$ (and not $(\hat{\mathcal{D}})_+$). Next, to obtain $\mathcal{E}$, we now discard not only diagrams $D \in \mathcal{D}$ admitting a wave move but also such that do so after a flype. (Our understanding is that we apply $\bar{t}'_2$ moves only on non-trivial $\sim$-equivalence classes, so that we may also use type A flypes.)

Regularization of V applies with $n = 6, 7, 8$, but we obtain now for $n = 7, 8$ a total of 330 diagrams E in $\mathcal{E}$ on which regularization fails. We can fix this, though, by restricting the condition (5.4) to diagrams $E \in \mathcal{E}$ (in our current notation) with $\sigma(E) \leq 0$. Then (5.4) is always satisfied. □

The calculation is now longer than for Proposition 5.2.1 due to the extra negative crossing. By (2.27) and semiadequacy, we also have restrictions when $g(D) \leq 4$. We left many of these tests out, though, adopting a simplistic attitude. As soon as the calculation became manageable (say, within a couple of days), we made very little effort to speed it further up by additional tests. We were aware that every single new step put into our procedure augments the risk of error. And when all diagrams can be verified, then certainly a subset can be, too, without that we become aware of overlooking the others. Indeed, a number of iterations were necessary to have the single steps working well and in the correct order.

5.6 On the number of unknotting Reidemeister moves

There has recently been some interest in the literature to estimate the minimal number of Reidemeister moves needed for turning an unknot diagram D of n crossings into the trivial (0-crossing) diagram (see e.g., [HL, HN, Hy]). Let us call this quantity here $r(D)$.

It was long known from Haken theory that $r(D)$ should be estimable from above, and in [HL] this was made explicit by showing that

$$r(D) \leq O(C^n), \tag{5.8}$$

where $n = c(D)$ (and $O(\dots)$ has the meaning from Corollary 4.2). Unfortunately, the exponential base C is a number with billions of digits.

This concerns only the *length* of the sequence. To bound the *number* of possible sequences, one must keep in mind that intermediate Reidemeister moves may augment the crossing number and, hence, also the number of applicable further Reidemeister moves. Thus, when $r(D) \leq N$, the number of sequences has, *a priori*, only a bound like $O((C_1N)^{C_2N})$, for some constants $C_1, C_2 > 0$, hyperexponential in N. This renders estimates like (5.8) quite impracticable. (There are, though, indeed attempts to use Haken theory computationally for this kind of algorithmical decision questions.)

Our above results imply the following statement, which can be regarded also as a continuation to Proposition 5.1.2:

Proposition 5.6.1 If D is an n crossing k-almost positive unknot diagram for $k \leq 4$ or k-almost special alternating unknot diagram for $k \leq 5$, then $r(D) = O(n^p)$, where p depends only on k.

This is in sharp contrast to the Hass–Lagarias bound in the general case. Because in Proposition 5.6.1 we consider only small k, we could also take p to be a global constant. Underscoring the dependence on k is made to indicate that our approach could work also for higher k, provided Proposition 5.2.1 or Theorem 5.5.4 can be extended. The values of p we will obtain for given k (the largest of which is 144) can likely be lowered by further direct calculation (as in Proposition 5.1.2 when $k \leq 2$). However, we opted for a more conceptual argument, formulating three lemmas below that have also some independent meaning.

For many diagrams D, we have $r(D) = O(n)$. Only recently some diagrams D were found where $r(D)$ is quadratic in n [HN]. Even although this still leaves a large gap between lower and upper bounds, we do not believe a polynomial upper bound is possible in general. We should note, though, that any set of (not-too-fancy) crossing number non-augmenting unknotting moves will yield much more tangible estimates in (5.8). For example, would Conjecture 5.4.4 be true, the below Lemmas 5.6.3 and 5.6.4, together with the result in [St19], would justify the constant $C = 10.399$ in (5.8). See also Remark 5.6.6.

Note also that, by Lemma 5.6.3, the difference between studying Reidemeister moves in S^2 or $\mathbb{R}^2$ is immaterial from the point of view of Proposition 5.6.1 (it affects p at most by ± 1).

Proposition 5.6.1 follows from the following three lemmas.

Lemma 5.6.2 For fixed k, the number of k-almost positive unknot diagrams of n crossings is polynomial in n.

Proof. There are $\binom{n}{k} = O(n^k)$ choices of negative crossings in an k-almost positive diagram once the underlying alternating diagram $\hat{D}$ is given.

Moreover $g = g(D) = g(\hat{D}) \le k$ if D is unknotted. Thus, by Theorem 3.1.1, D has at most $6g-3$ $\sim$-equivalence classes. Next, by Lemma 5.3.4, each $\sim$-equivalence class has at most g twist equivalence classes so that $\hat{D}$ has at most $g(6g-3)$ twist equivalence classes.

Thus, the number of $\hat{D}$ (regarded in S^2) of n crossings is at most $O(n^{g(6g-3)-1})$, with $g \le k$, and so the number of D is at most $O(n^{k(6k-3)-1+k})$. □

Lemma 5.6.3 A wave move in an n-crossing diagram can be realized by $O(n)$ Reidemeister moves. □

Lemma 5.6.4 A flype in an n-crossing diagram can be realized by $O(n^4)$ Reidemeister moves.

Proof. It is enough to prove the claim for the tangle isotopy in (a) below:

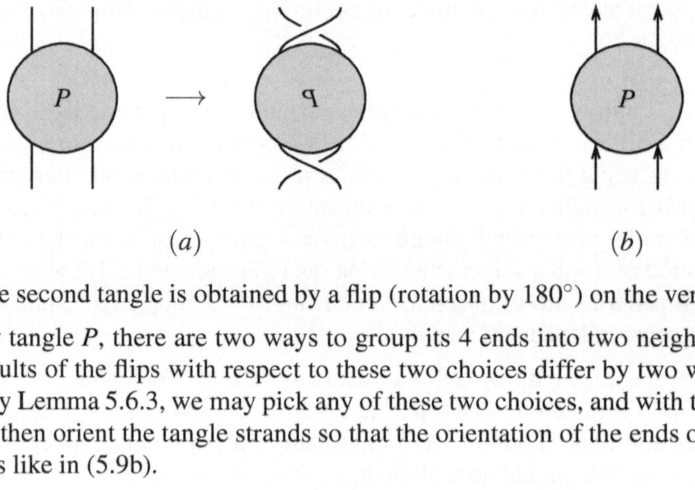

(5.9)

Here the second tangle is obtained by a flip (rotation by 180°) on the vertical axis.

For any tangle P, there are two ways to group its 4 ends into two neighbored pairs. The results of the flips with respect to these two choices differ by two wave moves. Thus, by Lemma 5.6.3, we may pick any of these two choices, and with this freedom we can then orient the tangle strands so that the orientation of the ends on the left of (5.9a) is like in (5.9b).

Now we would like to apply the procedure of Vogel [Vo] outside a trivial tangle (a) below,

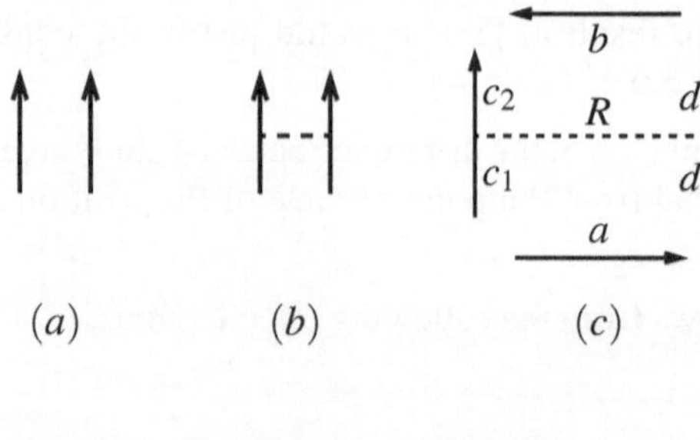

(5.10)

complementary to (5.9b), to obtain a diagram of P as a partially closed braid β. (We will use below the terminology of §2.9 for braids.)

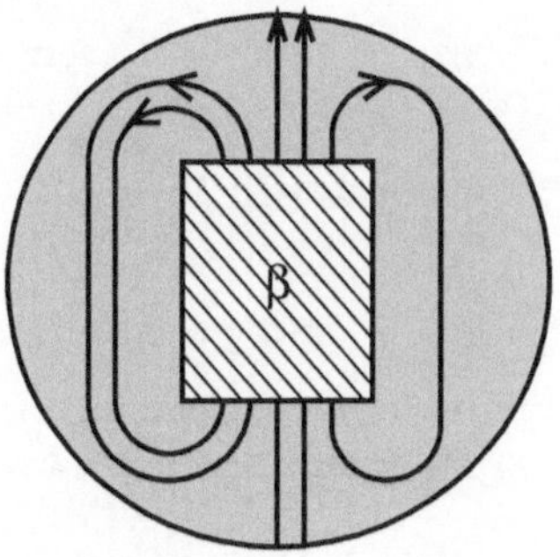

(5.11)

The Vogel move is a Reidemeister II move of a special type. It affects two edges in the boundary of a region R, which have the *same orientation* as seen from inside R, and belong to *distinct Seifert circles* and creates a reverse trivial clasp inside R.

To realize these moves in the tangle complementary to (5.10a), we draw a dashed line in (5.10a), as in (5.10b), and need to be concerned only with the Vogel moves that affect this line.

Let R be the region containing the line. The two arrows in (5.10b) are oriented oppositely as seen from inside R, and it is easy to see that for orientation reasons, these arrows must belong to distinct Seifert circles. If the pair (a,b) of segments in (5.10c) now allows for a Vogel move, then at least one of the four pairs (a,c_1), (a,d_1), (b,c_2) and (b,d_2) does also. So one can find a Vogel move that avoids the dashed line.

Vogel proves that, no matter what choice of move is taken at each step, after $O(n^2)$ applications no further moves are possible, and then the diagram is in braid form.

So we achieved the form (5.11) for P by $O(n^2)$ Reidemeister moves. Because the number of Seifert circles is not changed, β has $l = O(n)$ strands and $m = O(n^2)$ crossings. (All strands of β are closed except 2. There is no evidence how many strands are closed on the left and on the right.)

The reverse procedure can be applied after the flip with the same cost, so it is enough to look from now on at the case that P is in the form (5.11).

The flip of P now transcribes into the conjugation of β with the half-twist $\delta \in B_l$ (whose square is generating the center of B_l), plus the interchange of the strands closed on the left and on the right in (5.11). The latter step can be accomplished by $O(n^3)$ moves because by Lemma 5.6.3, moving one strand between either side takes $O(m) = O(n^2)$ moves.

It remains to look at the flip of the braid β, i.e., the conjugation with δ. With

$$\gamma_k = \sigma_1 \ldots \sigma_k \,,$$

we can write

$$\delta = \gamma_{l-1} \dots \gamma_1 . \tag{5.12}$$

To simplify things slightly, let us introduce the *bands*

$$\sigma_{i,j}^{\pm 1} = \sigma_i \dots \sigma_{j-1} \sigma_j^{\pm 1} \sigma_{j-1}^{-1} \dots \sigma_i^{-1} , \tag{5.13}$$

for $1 \le i < j \le l$, so that

$$\sigma_i = \sigma_{i,i+1} . \tag{5.14}$$

Let

$$\pi : B_l \to S_l , \qquad \pi(\sigma_i) = \tau_i = (i\, i+1)$$

be the permutation homomorphism (with S_l denoting the symmetric group and τ_i a transposition). It now is easy to check that, *unless* $i < k+1 < j$,

$$\gamma_k \sigma_{i,j} = \sigma_{\tau(i),\tau(j)} \gamma_k , \tag{5.15}$$

with $\tau = \pi(\gamma_k)^{-1}$, and that this identity is accomplished by $O(l)$ YB relations (recall §2.9).

Now, write β as a word in $\sigma_{i,j}$ using (5.14). Next, replace this word by $\delta^{-1}\delta\beta$ by $O(l^2) = O(n^2)$ Reidemeister II moves, using (5.12) to express δ. Then, by using (5.15), one can "pull" each γ_k through β (and the situation $i < k+1 < j$ never occurs!). Thus, by $O(m \cdot l) = O(n^3)$ applications of (5.15) one changes $\delta\beta$ to $\bar{\beta}\delta$ (with $\bar{\beta}$ being β with σ_i changed to σ_{l-i}), and this requires then $O(n^3 \cdot l) = O(n^4)$ Reidemeister moves.

After that, there are two half-twists of opposite sign on the left and right closed strands, which cancel by $O(n^2)$ moves, and the interchange of the left and right closed strands, which, as argued, costs $O(n^3)$ moves. □

Remark 5.6.5 It appears that the braid form is not essential, and in the statement of the lemma possibly $O(n^4)$ could be made into $O(n^3)$. It seems, though, very awkward to write a concrete procedure down without using braids.

Proof of Proposition 5.6.1. With Proposition 5.2.1, Theorem 5.5.4, and Lemmas 5.6.2 to 5.6.4, after switching between $O(n^{6k^2-2k-1})$ different k-almost positive unknot diagrams, each time using $O(n^4)$, so totally $O(n^{6k^2-2k+3})$ Reidemeister moves, we can reduce the crossing number. Then this procedure can be iterated. So we need at most $O(n^{6k^2-2k+4})$ moves, that is, we can set $p = 6k^2 - 2k + 4$ (for moves in S^2). □

Remark 5.6.6 An analogue of Proposition 5.6.1 holds for almost alternating diagrams by Theorem 2.6.3 and using Lemmas 5.6.3 and 5.6.4. Lemma 5.6.2 is not true for almost alternating diagrams, but its use can be avoided. While by flypes one can in general obtain exponentially many diagrams, only a polynomial number of flypes is needed to pass between any two diagrams. This can be seen from the structure of flyping circuits discussed in [ST] (see also the proof of Lemma 3.2.2).

5.7 Achiral knot classification

We turn, concluding this section, to achiral knots. The proof of Theorem 1.5.3 divides into two main cases, those for prime and for composite knots. For the prime case we have done the most work with the previous explanation. However, for composite knots some further arguments are necessary, the problem being that prime factors of achiral knots are not necessarily achiral. We will need some of the results in [St2] and Thistlethwaite's work [Th].

Proof of Theorem 1.5.3 for prime knots. Let D be a ≤ 4-almost positive diagram and K the achiral knot it represents. We will work by induction on the crossing number of D, showing that each time we can find a wave move to simplify D, unless D has small crossing number. (For at most 16 crossings the claim of the theorem was verified in [St6].) Now assuming that K is prime, we know that if D is composite, all but one of the prime factors of D represent the unknot. Using the result for the unknot we already proved, we can trivialize these factors, and then go over to deal with the non-trivial one. With this argument, we can w.l.o.g. assume that D is prime.

The proof now follows the one for Theorem 1.5.1 with the following modifications. To determine $\hat{D}$, the Rudolph–Bennequin shortcut is allowed. The inequality (2.6) must hold for an achiral knot K by replacing $g_s(K)$ by 0, even if the knot K is not slice. This is because one of $K\#K$ or $K\#-K$ is slice, and the estimate on the right is not sensitive w.r.t. orientation and additive under (proper) connected diagram sum. The σ tests in determining $\hat{D}$ are allowed because $\sigma = 0$ by achirality.

Lemma 5.1.3 shows that one can apply the semiadequacy shortcut if $c_-(D) = g(D)$ also here. So for such D, we are immediately done with the verification in the proof of Theorem 1.5.2. Even if $c_-(D) > g(D)$, the semiadequacy test is partly legitimized, namely if $X \in \mathcal{D}$ satisfies $\min\deg V(X) \geq -1$. This can be explained as follows from the work in [Th].

If X is B-adequate, then it has the minimal number of positive crossings, which is at most 4 since K is achiral and ≤ 4-almost positive. But X has also at most 4 negative crossings, so $c(X) \leq 8$, and we are easily out. If X is A-adequate, then positive $\bar{t}'_2$-twists preserve A-adequacy and $\min\deg V$. So $\min\deg V(D) \geq -1$, but $V(D)$ is reciprocal; if $\min\deg V \geq 0$, then $V = 1$ and contradiction to the semiadequacy. Then $\min\deg V(D) = -1$; moreover, $[V(D)]_{t^{-1}} = [V(D)]_{t^1} = \pm 1$ by semiadequacy. Then we use the property $V(1) = 1$ (see [J2, Section 12]), which shows that only two polynomials are possible, $-1/t + 3 - t$ and $1/t - 1 + t$. These are ruled out using $V(i) = \pm 1$ and $V(e^{\pi i/3}) \neq 0$.

Regularization (as described at the end of the proof of Theorem 1.5.1) applies similarly, and again we have $3 \leq n \leq 6$. Now $\max\deg V \neq 0$, but $\max\deg V \leq 6$, using that $\min\deg V \geq -6$ by [St6] and reciprocity of V. So regularity shows that we can deal with all diagrams D, unless $c(D) \leq n + \max\deg V \leq 12$. In that case, the check of [St6] for ≤ 16 crossings finishes the proof. □

We turn now to the two remaining knots. While $4_1\#4_1$ is easy to deal with inductively, we will need some work for $3_1\#!3_1$.

Lemma 5.7.1 Let K' be a chiral connected sum factor of a ≤ 4-almost positive achiral knot K. Then $\max\deg_z F(K') \leq 3$.

Proof. We apply (2.23), so in our case ≤ 4-almost positivity and achirality show $\max\deg_z F(K) \leq 7$. Now because K is achiral but K' is not, K must have $!K'$ as a factor too, i.e., $K = K'\#!K'\#L$, where L is achiral. As $\max\deg_z F(L) \geq 0$, we have $7 \geq \max\deg_z F(K) \geq 2\max\deg_z F(K')$, and so $\max\deg_z F(K') \leq 3$. □

Lemma 5.7.2 Let D' be a genus one diagram occurring as a prime factor of D. Assume D' represents a chiral knot K'. Then this knot K' is one of the trefoils.

Proof. By [St4], K' is a rational or pretzel knot, so it is a Montesinos knot. By [LT] then we have $\max\deg_z F(K') \geq c(K') - 2$. By the previous lemma $c(K') - 2 \leq \max\deg_z F(K') \leq 3$, so $c(K') \leq 5$. Now 4_1 is achiral, and the 5-crossing knots are excluded because $\max\deg_z F = 4$. So K' must be one of the trefoils. □

Proof of Theorem 1.5.3 for composite knots. Let D be a ≤ 4-almost positive diagram and K the composite achiral knot it represents. We have $g(D) \leq 4$. We assume w.l.o.g. that D cannot be further simplified by flypes and wave moves.

Let first D be prime. As D admits no wave move after flypes, the proof in the prime knot case shows that $c(D) \leq 12$. By Lemma 5.7.1, we have $\max\deg_z F \leq 7$. Then we see by check of the maximal degree of the F polynomial of achiral prime knots of at most 12 crossings that if K contains two achiral connected sum factors, then $K = 4_1\#4_1$. If K has a chiral factor K', then $K = K'\#!K'\#L$, where L is achiral. Then we use Lemma 5.7.1 and verify that only the trefoils have $\max\deg_z F(K') \leq 3$ among connected sum factors of knots of ≤ 12 crossings, and that L must be trivial.

So for the rest of the proof we assume that D is composite. If all prime factors of D are achiral, then again we can deal with these factors separately. With the argument in the preceding paragraph, D depicts some connected sum of trefoils and figure-8 knots. Because of $\max\deg_z F(D) \leq 7$, there remain only the options that we want to show. So assume below that at least one prime factor of D depicts a chiral knot.

Case 1. Let D have first at least 3 prime factors $D = D_1\#D_2\#D_3$. Because $4 \geq g(D) = \sum g(D_i)$, we may assume w.l.o.g. that $g(D_1) = g(D_2) = 1$. Then the knots $K_{1,2}$ these diagrams represent are trefoils (by Lemma 5.7.2). If they are same sign trefoils, then K must factor as $3_1\#3_1\#!3_1\#!3_1\#L$ for an achiral knot L, and $\max\deg_z F \geq 8$, a contradiction to $\max\deg_z F(K) \leq 7$. So $K_{1,2}$ are trefoils of opposite sign. Because $a_-(!3_1) = 3$ (see (2.21)), one of $D_{1,2}$ has at least three negative crossings. So D_3 is has at most one negative crossing but represents an achiral knot. We know, though, that non-trivial positive or almost positive knots are not achiral. So $K = 3_1\#!3_1$, and we are done.

Case 2. Now we consider the situation that $D = D_1\#D_2$ has two prime factors $D_{1,2}$.

Let $K_{1,2}$ be the knots of $D_{1,2}$. We can group the mirrored chiral prime factors of K which occur on different sides of the decomposition $D = D_1\#D_2$ into a (possibly composite) factor knot K' of K, and assume that $K_1 = K'\#L$ and $K_2 = !K'\#M$ with L and M achiral (and possibly trivial) and K' chiral (and non-trivial by the assumption preceding the case distinction). We have by Lemma 5.7.1 that $\max\deg_z F(K') \leq 3$.

Case 2.1. D_1 is positive, D_2 is ≤ 4-almost positive.

Case 2.1.1. $g(D_1) = 1$. Then $g(K'\#L) = 1$ so L is trivial, and K' is a positive knot of genus 1. By Lemma 5.7.2 it is the positive (right-hand) trefoil. Now consider D_2, with $g(D_2) \leq 3$.

Case 2.1.1.1. $g(D_2) = 1$. Then by Lemma 5.7.2, K_2 is a trefoil, and by achirality of K it must be negative, so we are done.

Case 2.1.1.2. $g(D_2) = 3$. Because D_2 is ≤ 4-almost positive, the left Morton inequality in (2.25) shows $\min\deg_l P(D_2) \geq -2$. Moreover D_1 is positive of genus 1, so $\min\deg_l P(D_1) = 2$. It follows that $\min\deg_l P(D) \geq 0$. Now K is achiral, and then [LM, Proposition 21] shows $P(D) = 1$. But the trefoil polynomial must divide $P(D)$, a contradiction.

Case 2.1.1.3. $g(D_2) = 2$. We repeat the argument in the previous case. Now $\min\deg_l P(D_2) \geq -4$, so $\min\deg_l P(D) \geq -2$. The case $\min\deg_l P(D) \geq 0$ is ruled out as before, so assume that $\min\deg_l P(D) = -2$, and then $\max\deg_l P(D) = 2$. Again $K = K'\#!K'\#M$, with M achiral. Comparison of the P-degrees and [LM, Proposition 21] shows then that $P_M = 1$. So we seek an achiral knot M with trivial skein polynomial whose connected sum with a negative trefoil $!K'$ is a knot K_2 in a ≤ 4-almost positive genus 2 diagram D_2. However, in that case $V(M) = 1$, so $\max\deg V(K_2) = -1$, and also $\sigma(K_2) < 0$. Then, assuming w.l.o.g. that we reduced D_2 by flypes and wave moves, we checked in the proof of Theorem 1.5.1 that $c(D_2) - \max\deg V(D_2) = n \leq 6$, so $c(D_2) \leq 5$. Then clearly M is trivial, and we are done.

Case 2.1.2. D_1 is positive and $g(D_1) = 2$. In the proof, given in [St2], of the fact that positive knots of genus 2 have minimal positive diagrams, we verified that all positive diagrams D_1 of genus 2 can be reduced to diagrams D_1' with $\max\deg Q(D_1') \geq c(D_1') - 2$. Now $\max\deg Q(D_1') \leq \max\deg_z F(D_1') \leq \max\deg_z F(D) \leq 7$. So $c(K_1) \leq c(D_1') \leq 9$, and K' is a chiral knot with $\max\deg_z F(K') \leq 3$ occurring as a factor of a knot K_1 with at most 9 crossings. Then again K' is a trefoil. But then, looking at the signature σ and using that L is achiral, we have $\sigma(D_1) = \sigma(K') + \sigma(L) = \pm 2$; while for a positive diagram D_1 of genus 2, we showed in [St2] that $\sigma = 4$. This contradiction finishes the case.

Case 2.1.3. D_1 is positive and $g(D_1) = 3$. So $g(D_2) = 1$ and by Lemma 5.7.2, K' is a trefoil (and M is trivial). Now again $\sigma(D_1) \geq 4$ by [St3], while $\sigma(D_2) = \sigma(K') = \pm 2$, and $\sigma(D) = \sigma(D_1) + \sigma(D_2) > 0$, which is a contradiction.

Case 2.2. D_1 is almost positive, D_2 is ≤ 3-almost positive.

Case 2.2.1. $g(D_2) = 3$. Then $g(D_1) = 1$ and K' is a trefoil, which must be right-hand because D_1 is almost positive. Using (2.25), we have $\min\deg_l P(D_2) \geq 0$, so $\min\deg_l P(D) > 0$ and a contradiction to achirality.

Case 2.2.2. $g(D_2) = 2$. By Morton's inequality $\min\deg_l P(D_2) \geq -2$. Also we know from [St2] that there is no almost positive knot of genus 1, so if $g(D_1) = 1$, then K_1 is positive. Then by Morton $\min\deg_l P(D_1) \geq 2$, and $\min\deg_l P(D) \geq 0$. So $P(D) = 1$. However, if $\min\deg_l P(D_1) \geq 2$, then $P(D_1)$, which divides $P(D)$, cannot be a unit in $\mathbb{Z}[l^{\pm 1}, m^{\pm 1}]$ by [LM, Proposition 21], and we have a contradiction.

Case 2.2.3. $g(D_2) = 1$. Then by Lemma 5.7.2, $!K'$ is a trefoil (and M is trivial). It must be a negative trefoil because otherwise $\sigma(D_2)$ and $\sigma(D_1)$ are both positive, and $\sigma(D) = \sigma(D_1) + \sigma(D_2) > 0$. So K' is a positive trefoil.

Case 2.2.3.1. If $g(D_1) = 1$, we are easily done by Lemma 5.7.2.

Case 2.2.3.2. If $g(D_1) = 3$, then by Morton's inequality $\min\deg_l P(D) \geq 0$, and we have a contradiction along the above lines in Case 2.1.1.2.

Case 2.2.3.3. So assume $g(D_1) = 2$. Now D_1 contains a positive trefoil factor. By Morton, $\min\deg_l P(D) \geq -2$, so by achirality and [LM, Proposition 21], and then along the lines in Case 2.1.1.3, we see that L must have trivial skein polynomial. So K_1 has the trefoil polynomial, and an almost positive diagram D_1. Then by [St5], we have $2 = \max\deg_m P(K_1) = 2g(K_1) = 2g(K') + 2g(L)$. (One can use alternatively Δ and the theorem that $\max\deg\Delta = 1 - \chi$ in [St14].) This implies that L must be trivial, and we are done.

Case 2.3. D_1 and D_2 are both 2-almost positive.

If $g(D_1) = 1$, then (L is trivial and) K' is a chiral knot with $\max\deg_z F \leq 3$ and a 2-almost positive genus 1 diagram. Up to reducible diagrams or diagrams with trivial clasps (which we can exclude) there are no such knots. Similarly, we argue if $g(D_2) = 1$. So assume $g(D_1) = g(D_2) = 2$. Then by Morton's inequality we have $\min\deg_l P(D_{1,2}) \geq 0$, and can obtain a contradiction in a way which should meanwhile be well known.

The case distinction and the proof are complete. □

The proof of Theorem 1.5.3 suggests that, beyond the description of the occurring knots, we can also obtain a partial simplification statement for their diagrams. We make this explicit in the following way.

Corollary 5.7.3 Let K be a k-almost positive achiral knot for $k \leq 4$, and $K \neq 3_1\#!3_1$. Then one can simplify a ≤ 4-almost positive diagram D of K by reducing wave moves and factor slides either to a minimal crossing diagram or to the 8 crossing diagram of $K = 6_3$ shown on the right.

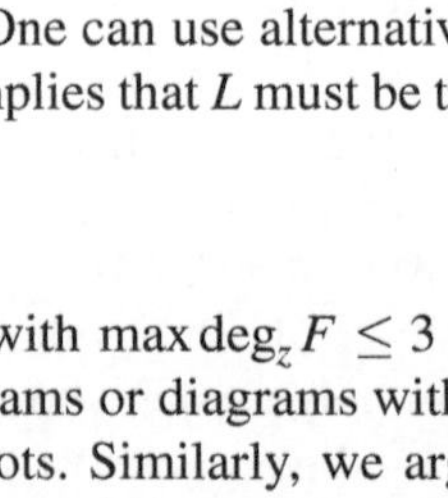

Proof. The exclusion of $K = 3_1\#!3_1$ was made to spare us the technical complications of dealing with non-amphicheiral prime factors. Thus, we can again consider w.l.o.g. prime diagrams.

Now the proof of Theorem 1.5.1 can be carried through with the only difference coming in the Jones polynomial regularization of $\mathcal{E}_n$ at the very end. (As before, the cases $c_-(D) \leq 2$ and $g(D) = 1$ can be checked directly.)

We can discard all alternating diagrams E in $\mathcal{E}$ because $\bar{t}'_2$-twists preserve alternation, and these are then minimal crossing diagrams as we claimed. From the rest we check that

$$\max\deg V(E) + \min\deg V(E) \geq -1\,. \tag{5.16}$$

(It is enough to consider only $E \in \mathcal{E}$ with $\max\deg V(E) \leq 6$.)

The property (5.16), together with the growth of $\max\deg V(E)$ under twists (which was the subject of regularization), implies that V will not be self-conjugate on diagrams that are obtained by positive $\bar{t}'_2$-twists from diagrams $E \in \mathcal{E}$.

It is thus enough to look at E themselves. We identify all diagrams that depict achiral knots, and this leads to the shown 8 crossing diagram of 6_3 (which admits no reducing wave move, and a few diagrams of $3_1\#!3_1$, which we chose not to dwell upon). □

From the work in §5.6 we have then also:

Corollary 5.7.4 Let K be a k-almost positive achiral knot for $k \leq 4$, and $K \neq 3_1\#!3_1$. Then one can simplify an n-crossing ≤ 4-almost positive diagram D of K to a minimal crossing diagram by $O(n^p)$ Reidemeister moves, where p depends only on k. □

Apart from the generator compilation, the most decisive reason for the success of such proofs is the easy computability of the invariants we use, primarily the signature σ and Jones polynomial V. It still took some days of work on a computer to complete the proofs. However, rather than thinking of such an approach as tedious, we feel that it should be considered paying high tribute to this efficiency, which makes it not too hard to evaluate the invariants on thousands of diagrams. (The most complicated ones we encountered have 35 crossings.) Recently, a lot of attention has been given to "relatives" of σ and V defined in terms of homology theory. The computability of these new invariants is still sufficiently difficult so that performing with them a procedure similar to ours seems, unfortunately, out of question for quite a while.

Chapter 6

The signature

As a last application related to positivity, we can finally settle a problem, initiated in [St7], and whose solution has been suspected for a while. First we have

Proposition 6.1 The knot 14_{45657} (depicted in figure 4 of [St7]) is the only positive knot of genus 4 with $\sigma = 4$. It has only one positive (reduced) diagram, its unique 14 crossing diagram.

This is the first *non-alternating* positive knot, which now is known to have only one positive diagram.

Proof. Let us seek a positive diagram of such a knot. Series of composite and special generators have $\sigma \geq 6$ and are clearly ruled out. Checking the σ of the positively crossing-switched non-special generators reveals that $\sigma = 8$ occurs 1,927,918 times, $\sigma = 6$ occurs 6662 times, and $\sigma = 4$ only once, for the positive diagram of 14_{45657}. Applying a $\bar{t}'_2$ twist at whatever crossing of this diagram gives $\sigma = 6$. □

In general, the problem of what is the minimal signature of positive knots of some genus is very difficult to study. We devoted to it a separate paper [St12]. Some of the theoretical thoughts there can be applied in practice here to prove

Theorem 6.2 The positive knots of $\sigma = 4$ are:

1) all genus 2 knots,

2) an infinite family of genus 3 knots, which is scarce, though, in the sense that asymptotically for $n \to \infty$ we have

$$\frac{\#\{K : K \text{ positive}, g(K) = 3, \sigma(K) = 4, c(K) = n\}}{\#\{K : K \text{ positive}, g(K) = 3, c(K) = n\}} = O\left(\frac{1}{n^{10}}\right), \qquad (6.1)$$

and

3) the knot 14_{45657} (of genus 4).

Proof. The genus 2 case is obvious, and let us first argue briefly about the estimate (6.1) for genus 3. First, by the result of [SV] we know that the number of special alternating knots of genus 3 and n crossings behaves asymptotically like a constant times n^{14}. That these are asymptotically dense in the set of positive genus 3 knots follows from the result of [SV] that maximal generators are special alternating, and an estimate in [St8], which shows that

$$c(D) - c(K) \leq 2g(K) - 1\,, \tag{6.2}$$

for a positive diagram D of a (positive) knot K. (See §10.4.) This argument determines the behavior of the denominator on the left of (6.1) for $n \to \infty$.

To estimate (asymptotically and from above) the numerator, we again use (6.2). This basically (up to a constant) allows one to go over from counting positive knots to counting positive diagrams. Now the positive diagrams of genus 3 with $\sigma = 4$ were described in Section 4 of [St7] (see the remarks following Proposition 4.1 there). We know that we have infinite degree of freedom to apply $\bar{t}'_2$-twists in at most 5 $\sim$-equivalence classes. Thus, the number of relevant diagrams of n crossings is $O(n^4)$, and the behavior of the numerator of (6.1) is also clarified.

With this argument, and Proposition 6.1, for the rest of the proof, we can assume we settled $g \leq 4$, so we want to show that there is no positive knot of genus $g \geq 5$ with $\sigma = 4$.

Let D be a positive knot diagram. Then we consider a sequence of diagrams $D = D_0, \dots, D_n = \bigcirc$, which is created as follows:

1) ("Generalized clasp resolution") If D_i has two equivalent crossings (i.e., two crossings that form a clasp, parallel or reverse, after flypes; see Definition 2.11.2) then we change one of these two crossings, apply a possible flype that turns the crossings into a trivial clasp, and resolve the clasp. (This is the move (17) of [St12].) If there is no flype necessary, nugatory crossings may occur after the clasp resolution. Then possibly remove these nugatory crossings to obtain D_{i+1}.
2) ("Shrinking a bridge") If D_i has no two equivalent crossings, then we choose a piece γ of the line of D_i (with no self-intersections) such that the endpoints of γ lie in neighbored regions of $D_i \setminus \gamma$. This means that after proper crossing changes on γ so that it becomes a bridge/tunnel, we can apply a wave move that shrinks it to a bridge/tunnel γ_1 to length 1 in D_{i+1}. We choose between bridge/tunnel so that the resulting single crossing of γ_1 is positive. Herein, we assume that the length of γ, which is the number of crossings it passes, is bigger than 1. We call such a curve *admissible*. Among all such possible admissible γ (we showed in [St12] that some always exist) we choose one of minimal length. If several minimal length bridge/tunnels γ are available, we choose γ so that $g(D_i) - g(D_{i+1})$ is minimal.

Now assume there is a D with $\sigma(D) \leq 4$ and $g(D) \geq 5$. We will derive a contradiction. By considering the proper diagrams of the sequence found for D, we may assume w.l.o.g. that $D_1 = D'$ has genus ≤ 4 and $\sigma(D') \leq 4$. (Note that always $\sigma(D_{i+1}) \leq \sigma(D_i)$.)

If D' differs from D by a clasp resolution, then $g(D) \leq g(D') + 1$. So the only option is that $g(D) = 5$ and $g(D') = 4$. Because $\sigma(D') \leq 4$, we checked that D' is the diagram of 14_{45657}. (We write below the knot for its positive diagram because the latter is unique.) Then D is obtained from D' by possibly creating first a certain number of nugatory crossings, and then a clasp (so that nugatory crossings become non-nugatory), and an optional flype (if there were no nugatory crossings). It is easy to see that the previously nugatory crossings can be switched in D so that the clasp to become one that does not require nugatory crossings to be added in D'. Thus, if we check all positive diagrams obtained from 14_{45657} by adding a clasp *without* nugatory crossings to have $\sigma = 6$, we are done. Now, if the genus remains 4, we already checked it. If $g = 5$, we have a prime positive diagram of 16 crossings. Thus, we seek a prime ([O]) positive knot of $g = 5$ and ≤ 16 crossings. If the knot is alternating, then by [Mu5] we have $\sigma = 10$, so consider only non-alternating knots. A pre-selection from the table of [HT] using the (necessary) skein polynomial condition $\min\deg_l P = \max\deg_m P = 10$, shows that all these knots have $\sigma \geq 6$.

For the rest of the argument, assume that D' is obtained from D by shrinking a bridge. We would like to show that this second move is needed only in very exceptional cases, and they do not occur here.

We proved in [St12] that D' has at most one $\sim$-equivalence class of ≥ 3 elements, and they are 3. Now we use the argument in that proof to show even stronger restrictions.

Lemma 6.3 (1) Assume D' has ≥ 2 disjoint clasps (meaning that the pairs of crossings involved are disjoint; the clasps may be parallel or reverse, but *not* up to flypes). Then $g(D) \leq g(D') + 1$.
(2) Assume D' has ≥ 3 disjoint clasps. Then D' cannot be obtained from a diagram D by shrinking a minimal bridge.

Proof. (1) There is at least one clasp a that does not contain the crossing of the shrunk bridge/tunnel γ_1. Because D has no equivalent crossings by assumption, the curve γ must pass through the clasp a. We argued in [St12] that it intersects the two edges of the clasp only once. Then D has an admissible curve of length 3, and shrinking this curve to a one crossing curve reduces the genus by at most 1. By choice of admissible curve, we have the first claim.

(2) There are at least two clasps that do not contain the crossing of the shrunk bridge/tunnel γ_1. Because D has no equivalent crossings by assumption, the curve γ must pass through both clasps. Then it intersects at least 4 edges, but we saw that at each clasp we have an admissible curve of length 3 in D. This is a contradiction to the minimality of the admissible curve. □

The rest is a simple electronic check. We determined the prime generators of genus 2 (they are 24) and 3 whose positification has $\sigma = 4$ (they are 13 and described in [St7]). Apply flypes to them, an optional $\bar{t}'_2$ twist, and again flypes. Then check that all the resulting diagrams have at least 2 disjoint clasps, and we see that $g(D) \leq 4$ by part (1) of the lemma, so are done. The composite generators are only of genus 2 and directly ruled out the same way. Likewise, the diagram of 14_{45657} has 4 clasps, and we apply part (2) of the lemma. This completes the proof of Theorem 6.2. □

Remark 6.4 We remarked in [St15] the relation of the sequence of moves we described to Taniyama's partial order. Indeed, one can apply the above sort of proof in that context. For example, one can show that 4_1 dominates all knots except connected sums of $(2,n)$-torus knots.

Proposition 6.5 If K is an almost positive knot of genus $g \geq 3$, then $\sigma \geq 4$.

Proof. If the almost positive diagram is composite, we can conclude the claim from the fact that $\sigma > 0$ when a knot is almost positive. So assume D is prime. Because $g(K) \geq 3$, also $g(D) \geq 3$. We make almost positive the 13 generators with $\sigma = 4$ and 14_{45657}. In the latter case always $\sigma = 4$, and in former case this is also true up to 8 diagrams of 10_{145}, which has $g = \sigma = 2$.

Now we build for these 8 almost positive diagrams E the set E_* as in (5.2). Again if the negative crossing is equivalent to a positive one, we can discard the case by assuming we work with a least crossing almost positive diagram or because the diagram becomes positive. We can also work with one diagram per generator because the signature is mutation invariant, and so is the genus of the knots if their mutated diagrams are almost positive, as can be concluded easily from the work in [St5]. Now we check that for all diagrams in all E_* we have $\sigma \geq 4$ or that $\max\deg_m P = 2$. The latter option shows by [St5] that $g = 2$, and the work there shows also that g will not change under $\bar{t}'_2$ twists at a (positive) $\sim$-equivalence class of more than 1 crossing (in an almost positive diagram). Because all diagrams we need to check are obtained from some diagram in E_* by twists at a $\sim$-equivalence class of more than 2 crossings, we see that if $\sigma = 2$, then $g = 2$. □

Corollary 6.6 If K is a 2-almost positive knot, then $\sigma > 0$ except if K is a (non-positive) twist knot.

Proof. If D is a 2-almost positive prime diagram of genus at least 3, then we can switch one more crossing in the above set of 8 diagrams of 10_{145} and check that $\sigma = 2$. It remains to deal with $g(D) \leq 2$, and this was done in Section 6 of [St2] (see remark 6.1 therein). The composite case D follows again easily from the prime one. □

Even though the check in Section 6 of [St2] is somewhat tedious, we must consider this proof as a considerable simplification of, and far more elegant than, the one in [St7]. It was the lack of such proof that was so troublesome while writing [St7].

Chapter 7

Braid index of alternating knots

7.1 Motivation and history

We tried to apply a variant of the regularization of Lemma 5.2.3 to the skein polynomial. Our motivation was to address the problem studied by Murasugi [Mu4] on the sharpness of the MWF inequality (2.24) on alternating links.

If we have equality in (2.24), then one can determine the braid index from the skein polynomial. Observably, this often occurs. However, the examples of unsharpness, albeit sporadic, are diversely distributed, and it seems very difficult to make meaningful statements as to nice classes on which the inequality would be sharp. It was a reasonable conjecture that it would be so on alternating links. Murasugi worked on this conjecture and proved that it holds if the link is rational or fibered alternating. His results remain the most noteworthy ones. Then he and Przytycki proved in [MP] the cases when $|\max \operatorname{cf} \Delta| < 4$ but found a counterexample of a 15 crossing 4-component link and an 18 crossing knot.

The extreme paucity of such examples (see Proposition 7.4.1) still suggests that sharpness statements are possible for large classes of alternating links. In that spirit, we show Theorem 1.5.4. Note that the Murasugi–Przytycki link counterexample is of genus 3 so that the theorem does not hold for links in this form. Their knot counterexample has genus 6.

To prove Theorem 1.5.4, we need first to have a good control on the MWF bound. We would like to have a growth in degree under $\bar{t}'_2$-twists, and we can formulate analogous initial conditions to the case of V that ensure inductively under the skein relation such degree growth. Originally, this enabled us to deal with genus 2 and, with considerable effort, genus 3. However, we always needed the initial two twist vectors for each $\sim$-equivalence class. The resulting tremendous growth of the number of generators and crossings in the initial diagrams made the case of genus 4 intractable, except in special cases.

Then Ohyama's paper [Oh] turned our attention to [MP]. There an efficient graph theoretic machinery was developed to study the MWF bound. With the help of this machinery we were able to considerably improve our work and deal with genus 4 completely, and this is what we describe in Chapter 7. (We will, however, unlikely be able to do so for genus 5.)

One should be warned that the work in [MP] contains a gap. Now there is an account published in [St20] discussing the gap and to what extent it can be fixed. We review some details below because later in Chapter 8 we will show how to modify the arguments to obtain a Bennequin surface. In summary, Murasugi–Przytycki's inequalities are true, but one must realize that their definition of index has not the proper geometric meaning. It simplifies the true transformation of the Seifert graph under their diagram move in a way that is *a priori* incorrect but (fortunately) *a posteriori* turns out to still give the right quantity.

7.2 Hidden Seifert circle problem

The gap in [MP] occurred in our study of the diagram move of Figure 8.2 of [MP]. Murasugi–Przytycki seem to assume that Figure 8.2 is the general case, but we will explain that it is not. And taking care of the missing case leads to a modified definition of index, which we call ind_0 (see Definition 7.3.4). Roughly speaking, the correction needed is that, in certain cases, some edges in $\mathrm{star}\, v$ are not contracted (cf. Definition 2.9.2). So Murasugi–Przytycki's diagram move just proves instead of (2.28) that

$$b(L) \leq s(D) - \mathrm{ind}_0(D)\,. \tag{7.1}$$

Then the question is how do $\mathrm{ind}\,(D)$ and $\mathrm{ind}_0(D)$ relate to each other. We will argue that

$$\mathrm{ind}\,(D) \leq \mathrm{ind}_0(D)\,, \tag{7.2}$$

which justifies (2.28) (and its applications in Murasugi-Przytycki's memoir). After we found this argument, we speculated, based on our computational evidence, whether in fact always

$$\mathrm{ind}\,(D) = \mathrm{ind}_0(D)\,. \tag{7.3}$$

Later, this was indeed established by Traczyk, who proved in [Tr2] the reverse inequality to (7.2). Still our (much more awkward) definition of index needs (at least temporary) treatment to prove (7.2) or (7.3) and fix the gap in [MP]. A minor modification of ind_0 will also be used for the Bennequin surfaces later.

Recall Definition 2.9.2. Now we must understand the move of Murasugi–Przytycki that corresponds to the choice of a simple edge e and the contraction of the star of v in G. (To set the record straight, we should say that this move was also considered, apparently simultaneously and independently, by Chalcraft [Ch], although merited there only with secondary attention. With this understanding, we will still refer to it below as the Murasugi–Przytycki move.) This move is shown in figure 8.2 of [MP]. Let D be the diagram before the move and D' the diagram resulting from it. Let us, for simplicity, identify an edge with its crossing and a vertex with its Seifert circle (see the remark above Proposition 2.9.3). In this language, the move of Murasugi–Przytycki eliminates one crossing, corresponding to e. The crossings of the other edges $e' \neq e$ incident to v do not disappear under the Murasugi–Przytycki move. Instead, they become in D' parts of join factors of $\Gamma(D')$ that correspond to a Murasugi summand on the opposite side of the modified Seifert circle. See the proof of Lemma 8.6 in [MP] and Figures 7.1 and 7.2.

The subtlety, which seems to have been overlooked in the proof of [MP], is illustrated in Figures 7.1 and 7.2. The Seifert circles adjacent to v may be nested in D in such a way that, by relaying the arc of v by the move, one does (and can) *not* go along *all* Seifert circles adjacent to v. In the Seifert graph $G' = \Gamma(D')$ of D', some of the edges incident to v in $G = \Gamma(D)$ may not enter, as written in the proof of Lemma 8.6 in [MP], into block components that are 2-vertex graphs (with a multiple edge).

Still we see that by contracting the star of v in $G = \Gamma(D)$, we obtain a graph $\tilde{G} = G/v$, which is a contraction of $G' = \Gamma(D')$. (We will later describe *exactly* how G' is constructed from G, but let us for the time being use the easier to obtain G/v instead.) Here, contraction of a graph means that $\tilde{G}$ is obtained from G' by contracting some (possibly several or no) edges, and we allow multiple edges in G' to be contracted (by doing so simultaneously with all simple edges they consist of).

More precisely, the difference between the block component of $\tilde{G}$ and G' is that in the last block component of G' in Figure 7.2 the star of v is contracted to obtain the block of $\tilde{G}$. So for the proof of Lemma 8.6 in [MP] and (2.28), we actually need the following lemma.

Lemma 7.2.1 If a graph H' is a contraction of H, then $\text{ind}(H') \leq \text{ind}(H)$.

This lemma can be proved easily by induction on the number of vertices, using the definition of the index. (The main point is that decontraction does not increase edge multiplicity.) *Still it should be understood that the contraction of a vertex is not fully correct as modeling the Murasugi–Przytycki diagram move.*

7.3 Modifying the index

In the following, we will need to define several "relatives" of Murasugi–Przytycki's index. This results from our desire first to fix the aforementioned error and then to keep track of the braided surfaces. We summarize the various indices in the scheme

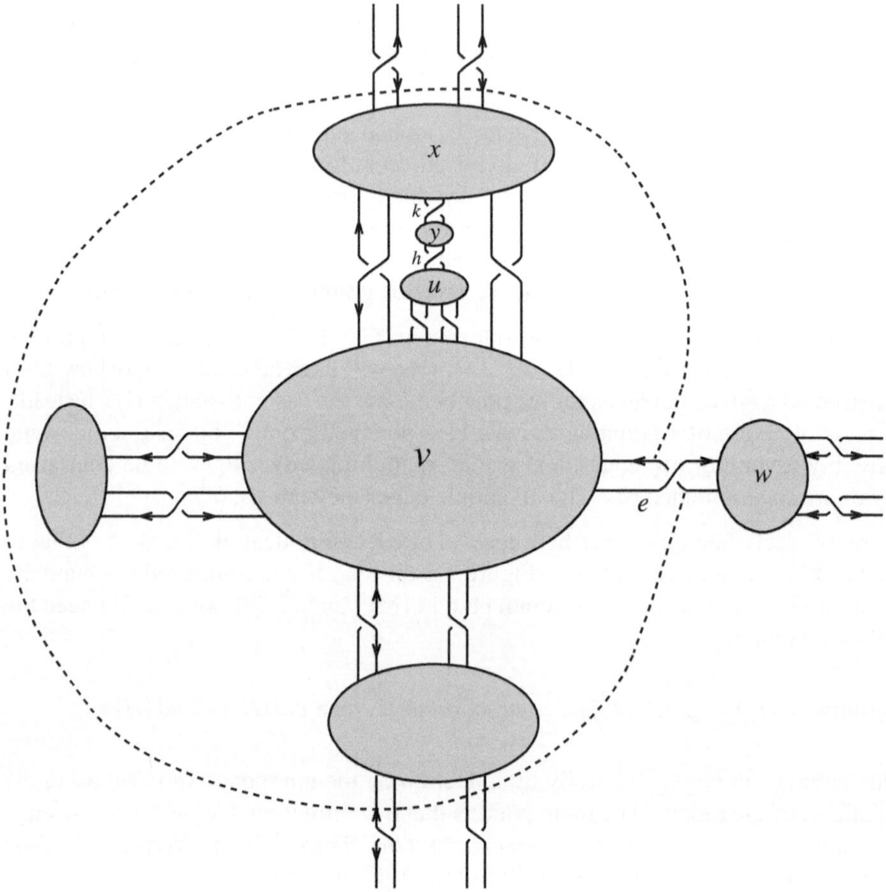

FIGURE 7.1: A move of Murasugi–Przytycki, where the relayed strand (dotted line) does not go along a Seifert circle (denoted as u) adjacent to v. The Seifert circles are depicted in gray to indicate that their interior may not be empty.

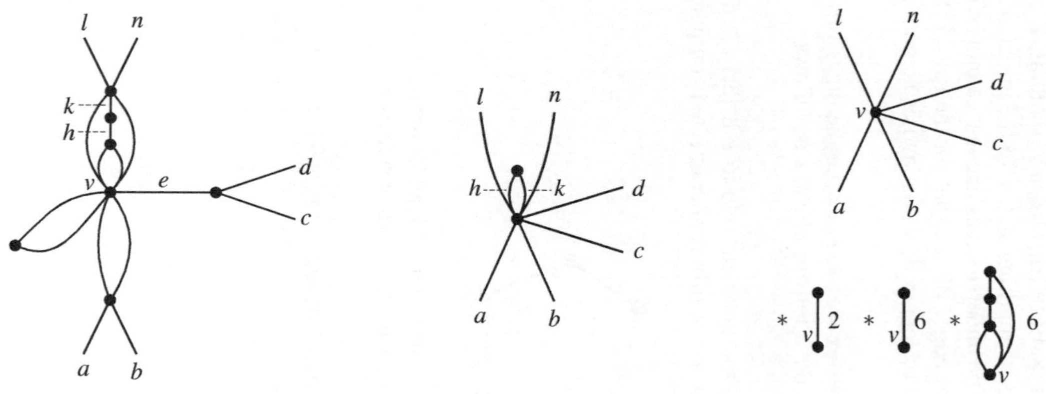

$G = \Gamma(D)$ $\quad$ $\tilde{G} = G/v$ $\quad$ $G' = \Gamma(D')$

FIGURE 7.2: The various Seifert graphs of the diagrams related to the move of Murasugi–Przytycki in Figure 7.1, in the case when the relayed strand does not go along all Seifert circles adjacent to v. The graph of D' is given in its block decomposition, which corresponds to the Murasugi sum decomposition along the newly created Seifert circle. For simplicity, we display a multiple edge by attaching the multiplicity to the edge drawn as simple (otherwise, a letter attached just indicates the name).

(8.4) and the remarks after it. The reader may consult this in advance to avoid confusion.

It becomes necessary to understand exactly the transformation of the Seifert graph G under the move of Murasugi–Przytycki. We describe it now and also fill in the detail overlooked by them.

Now we use graphs in which we mark edges. Each edge has a $\mathbb{Z}_2$-grading: it is either marked or not. The *marking* of an edge is its marking status. In the initial graph, all edges are unmarked. A marked edge is to be understood as one that cannot be chosen as an edge e. It corresponds to a multiple edge. *We assume* for the rest of the note that G *is bipartite*; thus G has no cycles of length 3, which avoids some technical difficulties.

We choose a non-marked edge e and a vertex v of e. Let w be the other vertex of e (see Figure 7.1). We define the notion *on the opposite side to e* as follows.

Definition 7.3.1 A vertex $y \neq v, w$ is on the opposite side to e if there is a vertex $x \neq v, w, y$ adjacent to v such that y and w are in different connected components of $(G \setminus v) \setminus x$.

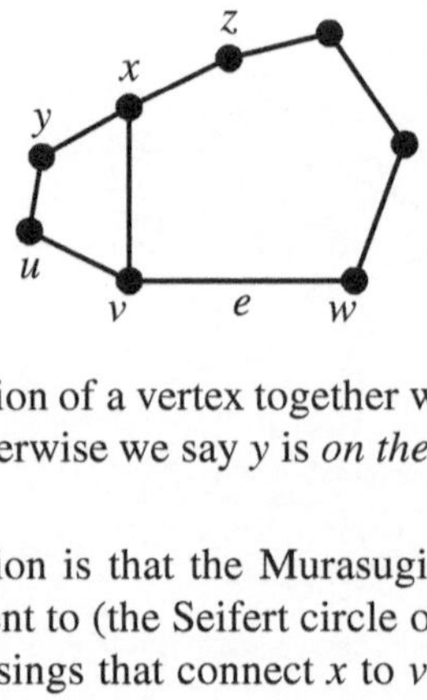

(7.4)

(Here "\" stands for the deletion of a vertex together with all its incident edges—but not its adjacent vertices.) Otherwise we say y is *on the same side as e*.

The meaning of this distinction is that the Murasugi–Przytycki move lays the arc along a Seifert circle x adjacent to (the Seifert circle of) v if x is on the same side as e. This move affects the crossings that connect x to v or to a Seifert circle z on the same side as e.

Definition 7.3.2 We define now the marked graph $G/_e v$. The vertices of $G/_e v$ are those of G except w. The edges and their markings are chosen by copying those in G as follows. Let an edge e' in G connect vertices $v_{1,2}$.

Case 1. v is among $v_{1,2}$, say $v = v_1$.

Case 1.1. If the other vertex v_2 of e' is w (i.e., $e = e'$), then e' is deleted.

Case 1.2. If v_2 is on the opposite side to e, then e' is retained in $G/_e v$ with the same marking.

Case 1.3. If v_2 is on the same side as e, then e' is retained in $G/_e v$ but marked.

Case 2. v is not among $v_{1,2}$.

Case 2.1. If none of $v_{1,2}$ is adjacent to v, then e' retains in $G/_e v$ the same vertices and marking.

Case 2.2. One of $v_{1,2}$, say v_1, is adjacent to v. (Then v_2 is not adjacent to v by bipartacy.)

Case 2.2.1. If $v_1 = w$, then change v_1 to v in $G/_e v$ and retain the marking.

Case 2.2.2. So assume next $v_1 \neq w$. If v_2 is on the opposite side to e, then retain $v_{1,2}$ and the marking.

Case 2.2.3. If v_2 is on the same side as e, then we change v_1 to v and retain the marking. (Note that by bipartacy, if v_2 is on the same side as e, then so must be v_1.)

Because being marked will indicate for us only that the edge cannot be chosen as e, the resulting graph $G/_e v$ may be reduced by turning a multiple edge into a single marked one. (This also makes it irrelevant to create multiple edges in Case 1.3.)

Theorem 7.3.3 If D' arises from D by a Murasugi–Przytycki diagram move at a crossing e and Seifert circle v, then $\Gamma(D') = \Gamma(D)/_e v$. □

Definition 7.3.4 Replace in Definition 2.9.2 the two occurrences of $G_{v_i} = G/v_i$ by $G/_e v_i$, as given in Definition 7.3.2. Then we define the corresponding notions of 0-*independent* edges and index $\text{ind}_0(D)$.

With this definition, we obtain (7.1). The property (7.2) can be proved by induction over the edge number, looking at a maximal set of independent (not 0-independent) edges, using Lemma 7.2.1 and the observation that G/v is a contraction of $G/_e v$. This fixes Murasugi–Przytycki's proof of Lemma 8.6 in [MP].

We speculated whether in fact (7.1) can be stronger than (2.28). If (7.3) is false, then Conjecture 2.9.5 is also false, so we were wondering whether this is a way to find counterexamples to that conjecture. We explained, though, that indeed (7.3) is true (and proved by Traczyk [Tr2] as a follow-up to our discovery of Murasugi–Przytycki's gap). One could then say that this makes our (much more awkward) definition of index obsolete. Still its treatment is necessary to prove (7.2) or (7.3) and to fix the gap in [MP].

Also, if one likes to keep the correspondence between (Seifert) graph and diagram, one must live with the circumstance that (in general) not all of star v is to be contracted. The idea of using vertex contraction (straightforwardly, following [MP]) appeared in at least one further paper, [MT]. We feel this misunderstanding may cause a problem at some point, which was additional motivation for us to provide the present correction.

We should stress that, while our proof of (7.2) might imply that independent edges (in Murasugi–Przytycki's sense of Definition 2.9.2) are 0-independent (i.e., corresponding to a set of single crossings admitting Seifert circle reducing operations), we *do not know* if the converse is true despite (7.3). Thus, proofs (like those we give below) relying on ind (rather than ind_0) are to some extent non-diagrammatic.

The important difference of ind_0 to ind lies in not affecting edges in Case 1.2. The treatment of vertices on the opposite side to e, the technical detail missed by Murasugi–Przytycki, does not affect the result by (7.3), yet it creates a lot of calculation overhead (which we experienced in attempts to use the possibly better estimate (7.1) prior to Traczyk's proof of (7.3)). Note, however, that it implies the additivity of ind_0 under block sum in an easier (and much more natural) way than Murasugi–Przytycki's corresponding statement for ind.

Definition 7.3.5 A marked graph is *not 2-connected* if it has an *unmarked* edge whose deletion disconnects it. If G is not 2-connected, there is a plane curve intersecting G in a single, and unmarked, edge. We call such a curve a *separating curve.*

We recall the following lemma from [St20].

Lemma 7.3.6 If $G_{1,2}$ are 2-connected, then $\text{ind}_0(G_1 * G_2) = \text{ind}_0(G_1) + \text{ind}_0(G_2)$.

7.4 Simplified regularization

To speed up the test for genus 4 generators, we make heavy use of Murasugi–Przytycki's work. We recall the inequalities (2.26)–(2.28) and the exactness of inequality (2.29) in alternating diagrams.

The proof of Theorem 1.5.4 will demonstrate the efficiency of Proposition 2.9.3 as a tool in determining the braid index. To give a first glimpse of that capacity, we mention that we have also obtained the following by computer verification of the tables in KnotScape [HT] and Knotilus [FR]:

Proposition 7.4.1 If K is an alternating knot of ≤ 18 crossings, then MWF is sharp, except if K is the (18 crossing) Murasugi–Przytycki knot or its mutant. □

This fact also shows the depth of the insight Murasugi–Przytycki must have had in picking up exactly these two knots as counterexample candidates!

The first easy lemma gives an upper control on the growth of the braid index under $\bar{t}'_2$ twists.

Lemma 7.4.2 If $\tilde{D}$ is obtained from D by a $\bar{t}'_2$-twist, then $\text{ind}(\tilde{D}) \geq \text{ind}(D) + 1$.

Proof. Let e be the edge in the Seifert graph $\Gamma(D)$ of D, which is bisected twice to obtain the Seifert graph $\Gamma(\tilde{D})$ of $\tilde{D}$. Double bisection means to put two valence-2-vertices $v_{1,2}$ on e, dividing it into three edges $e_{1,2,3}$. Let e_2 be the (middle) edge connecting $v_{1,2}$ in $\Gamma(\tilde{D})$.

For an independent edge set S in $\Gamma(D)$, we construct an independent edge set S' in $\Gamma(\tilde{D})$ by keeping $S \setminus \{e\}$ and including two of $e_{1,2,3}$ into S' if $e \in S$ and one of these three edges otherwise. This shows $\text{ind}(\tilde{D}) \geq \text{ind}(D) + 1$.

(If e connects a vertex of valence 2, then one easily sees that $\text{ind}(\tilde{D}) = \text{ind}(D) + 1$. Otherwise, we cannot exclude the possibility that $\text{ind}(\tilde{D}) = \text{ind}(D) + 2$, though.) □

Remark 7.4.3 Let us stress again that any choice of two respectively one edge(s) among $e_{1,2,3}$ to include into S' will do when $e \in S$ respectively $e \notin S$, provided we specify the to-be-contracted vertices correctly. This observation will be important both when $e \in S$ (for the proof of Corollary 7.4.5) and $e \notin S$ (for the proof of Lemma 7.4.4).

The next step is to control the degrees of the skein polynomial to estimate the braid index from below.

Lemma 7.4.4 Let D be an alternating diagram with (see (2.28))

$$\text{mwf}(D) = \text{mpb}(D). \tag{7.5}$$

Let e be an edge in the Seifert graph $\Gamma(D)$, which is not contained in at least one maximal independent set. (That is, there is a maximal independent set C with $e \notin C$.) Assume further that e is either a simple edge (i.e., its crossing has no Seifert equivalent one) or that D is special.

Let $\tilde{D}$ be obtained from D by a $\bar{t}'_2$ twist at (the crossing corresponding to) e. Then

1. (7.5) holds for $\tilde{D}$,
2. $\text{ind}(\tilde{D}) = \text{ind}(D) + 1$ and $\text{mwf}(\tilde{D}) = \text{mwf}(D) + 1$, and
3. e is not contained in some maximal independent set of $\Gamma(\tilde{D})$. (Here e is to be considered as a crossing in $\tilde{D}$ and identified with some of the three crossings after the twist.)

Proof. The equality (7.5), together with (2.26)–(2.28) and the exactness of inequality (2.29), imply that (2.26), (2.27) are exact for D. Now let D' be D with e switched. Assume w.l.o.g. e is positive in D (the negative case is handled analogously).

We want to show first

$$\min\deg_l P(\tilde{D}) = \min\deg_l P(D). \tag{7.6}$$

If D is special (alternating), (7.6) holds because both hand sides equal $1 - \chi(D)$. So assume e is simple. Then $\text{ind}_-(D') = \text{ind}_-(D) + 1$ because in $\Gamma(D')$ we have a single negative edge in a block. Denote by $q_1(D)$ and $q_2(D)$ the right-hand sides of the inequalities (2.26) and (2.27). Because $w(D') = w(D) - 2$, it follows that $q_2(D') = q_2(D)$, and since (2.27) is exact for D, we have

$$\min\deg_l P(D') \geq \min\deg_l P(D)\,. \tag{7.7}$$

Because (by the skein relation)

$$P(\tilde{D}) = (l^2+1)P(D) - l^2 P(D')\,, \tag{7.8}$$

we see that (7.6) holds.

Now let us look at (2.26). For an alternating diagram D, as remarked, all blocks of $G = \Gamma(D)$ have edges of the same sign. Thus, Theorem 2.9.4 implies that the intersection of each maximal independent set of G with the positive blocks also realizes $\text{ind}_+(G)$. The assumption that e is not contained in some maximal independent set of G (and the exactness of (2.29) in alternating diagrams) means then that $\text{ind}_+(D') \geq \text{ind}_+(D)$. Thus, $q_1(D') < q_1(D)$. Because (2.26) is exact for D, we see from (7.8) that

$$\max\deg_l P(\tilde{D}) = \max\deg_l P(D) + 2\,. \tag{7.9}$$

Now, (7.9) and (7.6) imply that $\text{mwf}(\tilde{D}) = \text{mwf}(D) + 1$, and because $s(\tilde{D}) = s(D) + 2$ and (7.5), we have that $\text{ind}(\tilde{D}) \leq \text{ind}(D) + 1$. The opposite inequality was shown in Lemma 7.4.2. So $\text{ind}(\tilde{D}) = \text{ind}(D) + 1$, and the second claim in our lemma follows. The first claim is then also clear. The third claim is a consequence of Remark 7.4.3.

□

Note that Condition 3 implies that one can iterate the twisting. So we have a test for sharpness of MWF on a given series.

Corollary 7.4.5 Assume D is a special alternating generator diagram such that (7.5) holds, and let S be the intersection of all maximal independent sets of D. Let for $S' \subset S$ the diagram $D_{S'}$ be obtained from D by applying one $\bar{t}'_2$ twist at each crossing in S'. Assume that for all S'

$$\text{mwf}(D_{S'}) = \text{mwf}(D) + |S'|\,. \tag{7.10}$$

Then the MWF inequality is exact on, and Conjecture 2.9.5 is true for the series of D.

Proof. As before, we split the series of D into $2^{|S|}$ subseries, depending on whether we twist or not at each crossing in S. We know that each twist augments the index by at least one. The equalities (7.5) and (7.10) then ensure that it goes up by exactly one under a $\bar{t}'_2$ twist at some choice of crossings s from S. So for each of the 3 $\sim$-equivalent crossings s' obtained from s under this twist, one can choose a maximal independent set not to contain s' by Remark 7.4.3. Then one can apply further $\bar{t}'_2$ twists at s, and the sharpness of MWF is preserved. □

For non-special generators, we must take into account Seifert equivalence classes.

Corollary 7.4.6 Assume for a non-special alternating generator diagram D that (7.5) holds, and let S, S' and $D_{S'}$ be as in Corollary 7.4.5. Let S'' be a set of crossings of D, such that the intersection of S'' with each Seifert equivalence class of n crossings contains at most $n-1$ elements. (In particular, S'' is disjoint from trivial Seifert equivalence classes, and so $S''\cap S=\varnothing$.) Assume that for all such sets S' and S'' we have

$$\mathrm{mwf}(D_{S'\cup S''}) = \mathrm{mwf}(D) + |S'| + |S''|\,. \tag{7.11}$$

Then the MWF inequality is exact on the series of D, and Conjecture 2.9.5 is true for this series.

Proof. The proof is completely analogous. When twisting further at the classes twisted in S'' (respectively the one remaining element in a Seifert equivalence class not in S''), all edges are simple. □

Note the following slight simplification. Each of the Seifert equivalence classes X_i of D decomposes by Lemma 2.11.3 completely into $\underset{*}{\sim}$-equivalence classes $Y_{i,j}$. If for two sets S'' and $\tilde{S}''$ we have $|S''\cap Y_{i,j}| = |\tilde{S}''\cap Y_{i,j}|$ for all applicable i,j, then $D_{S'\cup S''}$ and $D_{S'\cup \tilde{S}''}$ differ by a mutation. So we need to choose just one S'' for a given tuple of sizes $\{|S''\cap Y_{i,j}|\}_{i,j}$.

Proof of Theorem 1.5.4. This proof consists now in a verification that bases heavily on Corollaries 7.4.5 and 7.4.6. First consider one diagram D of each generator knot K.

The decisive merit of these corollaries is that they provide a condition to test sharpness of MWF by calculating P only on very few diagrams in the series and without adding a high number of crossings by $\bar{t}'_2$ twists. The previous verification needs $2^{t(D)}$ diagrams (recall Definition 2.11.5). Note that by Theorem 3.1.1 for $g(D)=4$, we have $c(D)\le 33$ and $t(D)\le 21$, so with that naive method, we would have up to 2^{21} diagrams per generator with up to 75 crossings.

The graph theoretic determination of S is not computationally trivial but takes only a fraction of the time that would have been required to calculate the large number of extra polynomials. It can be done by essentially the same recursion (and thus takes about as long) as the calculation of the index:

$$S = S_G = \bigcap\left\{\ \{e\}\cup S_{G/v}\ :\ \begin{array}{c} e \text{ simple edge of } G,\ v \text{ vertex} \\ \text{of } e,\ \mathrm{ind}(G/v) = \mathrm{ind}(G)-1 \end{array}\ \right\}.$$

(And of course $S=\varnothing$ if G has no simple edge.) This calculation reveals (see (7.12)) that S is always small: $|S|\le 5$. For most generators, S is actually empty.

For special generators the calculation of the index from the definition is more time consuming because they have more crossings than the non-special ones and admit no Murasugi-sum decomposition. This suggested for us to seek some further simplifications. We found an easy sufficient condition, which we describe in Proposition 8.3.2, to assure that S is empty. In case of a special generator D with empty S, only the polynomial of D needs to be calculated.

Proposition 8.3.2 is stated and proved in Chapter 8 because it relates to some further notions treated there. The generator 12_{1202} of genus 2 easily shows that the inequality (8.2) in the proposition is false for non-special generators. On the other hand, (8.3) holds, and (8.2) becomes an equality for all special genus 2 generators. For genus 4, it is still exact for slightly more than one half (748,193) of the special generators.

Even when Proposition 8.3.2 does not apply directly, its proof shows some priority that should be given to the contraction of vertices of valence two. Note that the only difficulty for non-special generators is that F may not be a forest, and we need to take care of not contracting vertices of valence 2 incident to a double edge. With this additional restriction, one can deal with non-special generators, too, and this considerably speeds up their calculation.

The following table shows the sizes of intersections S of maximal independent sets for special and non-special genus 4 generators; the last column gives the total number of extra, non-generator, knots whose polynomials needed to be calculated. The most complicated such knots have 35 crossings. For a special generator, their number is $2^{\#S} - 1$.

$\#S$	0	1	2	3	4	5	total test knots
special	1,082,270	269,406	106,204	20,676	1649	33	758,508
non-special	1,573,426	290,668	63,821	6,488	178	—	15,331,751

(7.12)

Even though the combination of S' with the sets S'' creates extra work for the non-special generators, this overhead was manageable. We used the simplification for the choice of S'' explained after the proof of Corollary 7.4.6. (There are theoretic ways to further reduce the calculation, but they result in too technical conditions, whose implementation augments the risk of an error.) In total, non-special generators required less time to deal with because the calculation of the sets S for special generators took about 15 times as long as for the non-special ones.

With this, however, the work is not yet finished. The lack of (at least confirmed) flype invariance of ind (D) requires us to deal with diagrams D_0 obtained by flypes from the generator diagrams D we considered. To simplify the occurring problem, first note that type B flypes (see Figure 2.6) commute with $\bar{t}_2'$ twists, so it is enough to generate D_0 from D by type A flypes. (This leads already only to relatively few new diagrams D_0, about 58,000 for special generators K and about 150,000 for non-special ones.) Also, the condition (8.3) is invariant under flypes, so no D_0 coming from diagrams D we discarded using Proposition 8.3.2 need to be considered. Moreover, the mutation invariance of P ensures that the MWF bound will grow (by 1) for each $\bar{t}_2'$ twist we apply on D_0. Because we argued that the diagram index will grow also (at least) by 1, the verification we need to perform on the remaining D_0 reduces to the check of (7.5). (We determined in fact also the intersections S of maximal independent sets for all the diagrams D_0 and found that they have the same size as for the corresponding diagrams D that differ by type A flypes.)

We must at last admit that, even after these various arguments and simplifications, still several weeks were necessary to complete the work. □

Remark 7.4.7 It is clear from our proofs that the Conjecture 2.9.5 of Murasugi–Przytycki is also confirmed for (at least one alternating diagram of) the knots in Theorem 1.5.4 and Proposition 7.4.1.

Using [STV], we have for example:

Corollary 7.4.8 The number of alternating genus 2 respectively 3 knots of braid index n grows like $O(n^8)$ respectively $O(n^{14})$. The number of achiral such knots grows like $O(n^2)$ respectively $O(n^5)$.

Proof. Regularity of all generators implies that, up to an additive constant, the braid index behaves like half of the crossing number on such knots. Then the results on enumeration by crossing number extend directly by replacing crossing number by braid index. □

7.5 A conjecture

The above work suggests that for a fixed genus, b should behave similarly to $c/2$. A calculation of the range of the difference $c-2b$ for alternating knots K of genus $g \le 4$ reveals that we have

$$g-2 \overset{(*)}{\le} c-2b \le 2g-3 \qquad \text{if } K \text{ is special alternating,} \tag{7.13}$$

$$-2 \le c-2b \le 2g-4 \qquad \text{if } K \text{ is not special alternating,} \tag{7.14}$$

with all inequalities realized sharply. In fact, both estimates from above are shown in [St13] for all alternating *knots* K (of all g), and we see that they are the best possible (at least for $g \le 4$). The lower estimate in (7.14) follows by the result of Ohyama [Oh]. The improved lower bound (*) in (7.13) seems unclear in contrast. It would be implied by the inequality

$$4\,\mathrm{ind}(G)+e(G) \ge 3(v(G)-1)\,, \tag{7.15}$$

for a (planar bipartite) graph G with $e(G)$ edges and $v(G)$ vertices and with *an odd number of spanning trees*. The spanning tree condition is necessary, as shown in simple examples, and its use makes an approach to prove (7.15) difficult. Still, we verified (*) and (7.15) for (the Seifert graphs of alternating diagrams of) all special alternating knots of up to 18 crossings and found no counterexample. If it holds,

(7.15) would then in fact imply (*) more generally for alternating knots all of whose Murasugi atoms (of the alternating diagram; see above Theorem 2.9.4) are knots (and none are links).

One could conjecture (*) also for links:

Conjecture 7.5.1 For a special alternating link L we have $c(L) - g(L) \geq 2(b(L) - 1)$.

In the case of links, (7.15) will take its extended form for general (planar bipartite) graphs G

$$4\,\mathrm{ind}\,(G) + e(G) \geq 3v(G) - 2 - n(D), \tag{7.16}$$

where D is the (special alternating) diagram with $\Gamma(D) = G$ and $n(D)$ the number of its components. (Note that G has odd number of spanning trees iff $n(D) = 1$; see [MS].) The other 3 inequalities in (7.13) and (7.14) hold also for links, but the upper estimates need some changes depending on $n(D)$.

There is one more important special case in which one can show our conjecture.

Proposition 7.5.2 If L is a special alternating arborescent link (*without* hidden Conway spheres), then $c(L) - g(L) \geq 2(b(L) - 1)$.

Remark 7.5.3 The hidden Conway spheres of Menasco [Me] refer to the links that are arborescent and alternating (i.e., have an arborescent diagram and an alternating diagram) but not alternating*ly* arborescent (i.e., do not have a diagram which is *simultaneously* arborescent and alternating). Even though this family of exceptions is relatively small, it exists and is often carelessly overlooked by authors. See §2.6 or [Th2] for an explanation.

Proof. We show (7.16) for a special alternating arborescent diagram D and $G = \Gamma(D)$. For a special diagram D, the Seifert graph G coincides with one of the checkerboard graphs (see Remark 2.9.6). If D is arborescent, the latter are series-parallel (§2.5.2). So we now assume G is (bipartite and) series-parallel. We use induction over the number $e_s(G)$ of simple edges of G. (Note that multiple edges enter into the count $e(G)$ with their multiplicity.)

First assume $e_s(G) = 0$. Because decreasing by one the multiplicity > 2 of a multiple edge changes $n(D)$ by ± 1, it is enough to deal with the case that all edges of G are double. Next, one can successively delete double edges (which preserves $n(D)$ and $\mathrm{ind}\,(G) = 0$) until one has a tree of double edges. Then the inequality (7.16) we claimed is easy to see (as an equality, with D being a connected sum of Hopf link diagrams).

Now assume $e_s(G) > 0$. We consider a simple edge e of G, which is the first in a maximal independent sequence.

This edge e can, as a first case, be adjacent to a vertex v connecting only two vertices

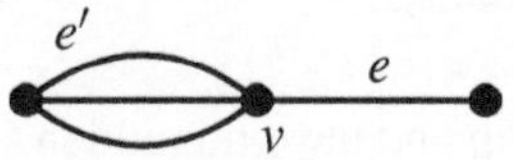

. (7.17)

We want to use the definition of index and contract v.

If e' is simple, then the contraction preserves $n(D)$ (it trivializes a reverse clasp) and reduces $v(G)$ by 2 and $\mathrm{ind}(G)$ at least by 1. The resulting graph is either series-parallel or a block sum of such. So again we can argue by induction and the stability of (7.16) under connected/block sum. If e' is multiple, then this can be easily reduced to the argument when e' is simple. This deals with the case (7.17).

If we have no fragment (7.17), then we have a simple edge e of the following form:

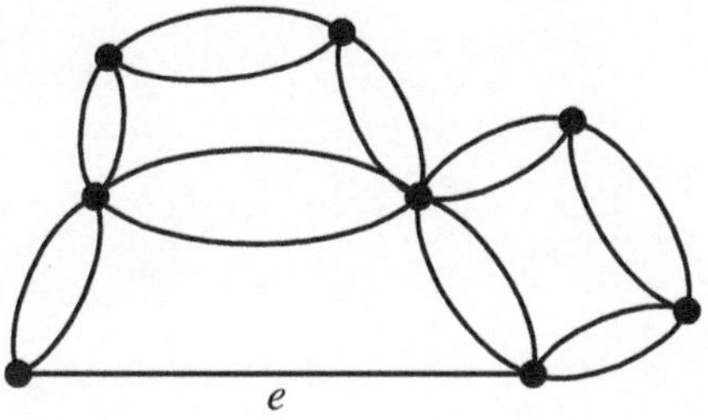

(7.18)

The edges except e are all multiple, but they can be reduced to double ones using the previous argument. Now from (7.18) one sees that there is always a cycle in which only two vertices are connected (not necessarily by simple edges, as shown below) to vertices outside the cycle.

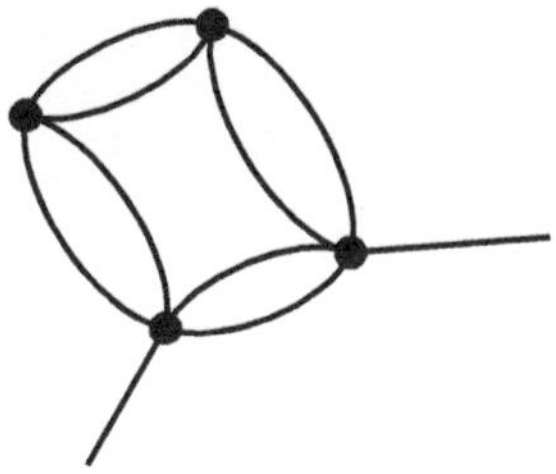

By contractions of two double edges like

we can lose 2 vertices, 4 edges, and $n(D)$ goes down by 2. The index does not go up by Lemma 7.2.1. It is clear that by such contractions and reducing a multiplicity > 3 of edges by 2, one can turn (7.18) into 

. (Keep in mind that the graph is bipartite.) Then e is no longer simple, and again we can use induction. □

Corollary 7.5.4 The inequality $c(L) - g(L) \geq 2(b(L) - 1)$ holds for special alternating Montesinos links L.

Proof. Such links are arborescent and have no hidden Conway spheres. □

Chapter 8

Minimal string Bennequin surfaces

8.1 Statement of result

We apply the work in the previous chapter to the problem of what knots have a braided surface of minimal genus on the minimal number of strands. With the notation of §2.9, let $\sigma_{i,k} \in B_n$ for $1 \leq i < k \leq n$ be the bands we introduced in (5.13):

$$\sigma_{i,k} := \sigma_i \ldots \sigma_{k-2}\sigma_{k-1}\sigma_{k-2}^{-1} \ldots \sigma_i^{-1}\,.$$

If one represents a link L as the closure of a braid $\beta \in B_n$, which is written as a product of l of the $\sigma_{i,k}$ and their inverses, one obtains a Seifert surface S of L consisting of n disks and l bands. We call S a *braided surface* of L. If S has minimal genus (i.e., equal to the genus of L), then S is called a *Bennequin surface*. This terminology was coined by Birman and Menasco (see for example [BM]), and it relates to the work of Bennequin. He showed in [Be] that such a surface exists for 3-braid links on a 3-string braid. Rudolph [Ru2] showed that *every* (not necessarily minimal genus) Seifert surface can be made into a braided surface. In particular, a Bennequin surface always exists on a braid of some (possibly very large) number of strings. A natural question was whether the minimal number of strings (i.e., the braid index of L) are always enough to span a Bennequin surface. With M. Hirasawa [HS], we showed later that there are knots of genus 3 and braid index 4, which have no minimal (i.e., 4-)string Bennequin surface. Hirasawa has also shown, contrarily, that for 2-bridge links *any* minimal genus surface is a braided surface on the minimal number of strands. (The same holds for 3-braid links because in [St13] it was shown that such links have a single minimal genus surface, in fact even just a single incompressible one.) Here we want to show the following result.

Theorem 8.1.1 Any alternating knot of genus up to 4 or at most 18 crossings has a minimal string Bennequin surface.

8.2 The restricted index

We now define a third variant of graph index, which keeps track of surfaces. As for ind_0 in §7.3, we use marked edges. In the initial graph $G = \Gamma(D)$, all edges are unmarked. A marked edge is to be understood as one that cannot be chosen as an edge e. It corresponds to a crossing that is grouped with other crossings to form a band or that connects the same 2 Seifert circles as some other crossing. *We assume* again for the rest of the treatment that *G is bipartite*, and thus has no cycles of length 3, which would create some problems.

First, we *reduce* the graph G by turning a multiple edge into a simple marked one.

Next we choose a non-marked edge e and a vertex v. Let w be the other vertex of e (see Figure 7.1). Recall the notion on the opposite side to e from Definition 7.3.1. The meaning of the distinction between the side of e and the opposite side was that the Murasugi–Przytycki move lays the arc along a Seifert circle x adjacent to (the Seifert circle of) v if x is on the same side as e. This move affects the crossings (possibly in bands) that connect x to v or to a Seifert circle z on the same side as e.

Definition 8.2.1 We now define the marked graph $G//_e v$. The vertices of $G//_e v$ are those of G except w. The edges and their markings are chosen by copying those in G as follows. Let an edge e' in G connect vertices $v_{1,2}$.

Case 1. v is among $v_{1,2}$, say $v = v_1$.

Case 1.1. If the other vertex v_2 of e' is w (i.e., $e = e'$), then e' is deleted.

Case 1.2. If v_2 is on the opposite side to e, then e' is retained in $G//_e v$ with the same marking.

Case 1.3. If v_2 is on the same side as e, then e' is retained in $G//_e v$ but marked.

Case 2. v is not among $v_{1,2}$.

Case 2.1. If none of $v_{1,2}$ is adjacent to v, then e' retains in $G//_e v$ the same vertices and marking.

Case 2.2. One of $v_{1,2}$, say v_1, is adjacent to v. (Then v_2 is not adjacent to v by bipartacy.)

Case 2.2.1. If $v_1 = w$, then change v_1 to v in $G//_e v$ and retain the marking.

Case 2.2.2. So assume next $v_1 \neq w$. If v_2 is on the opposite side to e, then retain $v_{1,2}$ and the marking.

Case 2.2.3. If v_2 is on the same side as e, then we change v_1 to v and make the edge marked. (Note that by bipartacy, if v_2 is on the same side as e, then so must v_1.)

Because being marked will indicate for us only that the edge cannot be chosen as e, the resulting graph $G//_e v$ may be again reduced by turning a multiple edge into a single marked one. (This also makes it irrelevant to create multiple edges in Case 1.3.)

Definition 8.2.2 We define the *restricted index* $\operatorname{ind}_b(G)$ of a graph G like the (Murasugi–Przytycki) index $\operatorname{ind}(G)$ in Definition 2.9.2, replacing G/v by $G//_e v$. Again, set $\operatorname{ind}_b(D) = \operatorname{ind}_b(\Gamma(D))$ for a diagram D.

Remark 8.2.3 If in Case 2.2.3 of the definition of $G//_e v$, we retain the old marking (and do not necessarily turn the edge marked), and then we ignore the restriction coming from keeping bands. So marked edges become the equivalent of multiple ones. Thus, the corresponding index is exactly the previously defined ind_0, correcting Murasugi–Przytycki's definition of index to reflect their diagram move.

Again we have

Lemma 8.2.4 If $G_{1,2}$ are 2-connected, then $\operatorname{ind}_b(G_1 * G_2) = \operatorname{ind}_b(G_1) + \operatorname{ind}_b(G_2)$.

Proof. Similar to Lemma 7.3.6. □

Proposition 8.2.5 Let D be a minimal genus diagram of a link L (i.e., the canonical surface of D is a minimal genus surface of L). Then L has a Bennequin surface on a braid of $s(D) - \operatorname{ind}_b(D)$ strings.

Proof. We call a *band* a set of crossings in a diagram which looks locally like a braid of the form

$$\sigma_1 \dots \sigma_{k-1} \sigma_n^{-1} \dots \sigma_{k+1}^{-1} \sigma_k^{\pm 1} \sigma_{k+1} \dots \sigma_n \sigma_{k-1}^{-1} \dots \sigma_1^{-1}$$

for some $1 \leq k \leq n$. (This is a fragment of a diagram that contains parts of $n+1$ Seifert circles.)

A marked edge corresponds to a crossing that either belongs to a band (of > 1 crossing) or connects the same Seifert circles as at least one other crossing. With this understanding, one has to verify that the definition of ind_b using $G//_e v$ reflects precisely the effect on the diagram by the Murasugi–Przytycki move, keeping track of markings. (Note, however, that a contraction at v can delete marked edges incident to v.)

Also, each Murasugi–Przytycki move acts within a single block. So we see that, starting with D, we can apply *at least* $\operatorname{ind}_b(D)$ Murasugi–Przytycki moves, and each move reduces the number of Seifert circles and bands by 1. Finally, the moves of Yamada [Ya] can likewise be applied so that the number of (Seifert circles and) bands is preserved. Then we have a braid diagram of L, which gives the desired surface. With this Proposition 8.2.5 is proved. □

8.3 Finding a minimal string Bennequin surface

It is clear now that with Proposition 8.2.5 we have a tool for a computational verification of Theorem 8.1.1. Unfortunately, the description of ind_b makes its calculation very time consuming. In practice, we tried to further speed up the recursion. We feel these algorithmical details are too technical and too long to present here, but roughly speaking, we tried to take advantage of the observation that handling of vertices "far apart" in the graph does not depend on their order. We still could not avoid that some sequences of edges were left unverified so that we may obtain smaller values than ind_b. But this computation was much faster, and when its bound already coincides with the one of MWF, we know that the one of ind_b must also coincide. This dealt with most generators much more quickly than ind_b would have done, and we had to calculate ind_b (following the definition) only for the handful of remaining ones.

We explain first the following partial case.

Proposition 8.3.1 Alternating knots of genus ≤ 3 have a minimal string Bennequin surface.

Proof. First, one checks (by computer) for all generators K and all their diagrams D the property

$$\text{ind}(D) = \text{ind}_b(D). \tag{8.1}$$

It is helpful to distinguish again flypes of type A and type B as in Figure 2.6. Flypes of type B commute with $\bar{t}'_2$ twists and type A flypes. So it is enough to generate the diagrams obtainable from a given diagram of a generator K by type A flypes. For each such diagram D, it is enough to find a diagram differing from D by type B flypes, which satisfies (8.1). The following argument shows that (8.1) is preserved under $\bar{t}'_2$ twists.

When (8.1) holds, then it is easy to see, either by a graph theoretic argument or by deforming the canonical surface geometrically, that a $\bar{t}'_2$ move at a crossing not deleted by a Murasugi–Przytycki move adds one disk and one band. Consider the case the crossing we twist at is deleted by a Murasugi–Przytycki move. So the edge e is in a maximal independent set, and we chose a vertex v to contract. Then we first apply a move

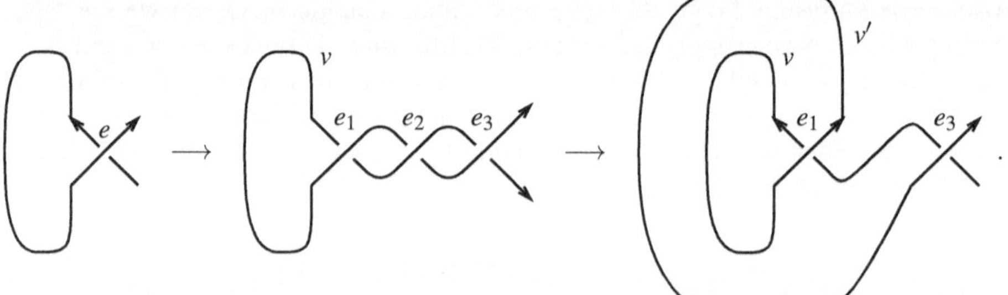

(In the notation, we identified edges with crossings and vertices with Seifert circles.) So we alter the middle of the 3 crossings after the $\bar{t}'_2$ move. (We lay the rest of the new Seifert circle v' below all other crossings attached to v from outside.) We now obtain at the site of the $\bar{t}'_2$ twist locally a braid $\sigma_2\sigma_1$. Then delete by a Murasugi–Przytycki move the edge e_3, contracting (the vertex of) v'. So the result follows. □

The extension to genus 4 requires both more calculation and more delicate implementation techniques of the index. Let us state, however, a simple and useful test that was already quoted in §7.4.

Proposition 8.3.2 Assume D is a special knot generator diagram, and $t(D)$ is the number of $\sim$-equivalence classes of D. Then

$$\mathrm{mpb}(D) = s(D) - \mathrm{ind}(D) \leq t(D) + \chi(D)\,. \tag{8.2}$$

Also, $\mathrm{ind}_b(D) \geq c(D) - t(D)$. Moreover, the intersection of all independent sets of size $c(D) - t(D)$ is empty. In particular, if

$$t(D) + \chi(D) = \mathrm{mwf}(D)\,, \tag{8.3}$$

then MWF is exact, and Conjecture 2.9.5 is true on the series of D.

Proof. Up to flypes assume that each non-trivial $\sim$-equivalence class of D forms a clasp. The diagram D has $c(D)$ crossings and $c(D) + \chi(D)$ Seifert circles. There are $c(D) - t(D)$ Seifert circles of valence 2. We claim that for each such Seifert circle we can take one of the edges adjacent to it into an independent set.

Now, because D is a knot diagram and is special, the edges incident to valence-2 vertices in the Sefiert graph Γ of D form a graph F, which is a forest. Moreover, for each region R of $\mathbb{R}^2 \setminus \Gamma$ there are at least two edges in $\Gamma \cap \partial R$ not incident to a valence-2 vertex. (There is at least one such edge because D is a knot diagram, and then at least one other because Γ is bipartite.) So one can contract all valence-2 vertices, obtaining $c(D) + \chi(D) - (c(D) - t(D)) = t(D) + \chi(D)$ vertices.

To find two disjoint independent sets, just choose the opposite edge at each valence-2 vertex. To see that $\mathrm{ind}_b(D) \geq c(D) - t(D)$, choose a root r in each tree of F, which is not of valence 2. Include in the independent set an edge that is incident to a valence-2 vertex v adjacent to r but not connecting r and v. Then contract v and proceed inductively. □

Proof of Theorem 8.1.1. For genus 4, again we need to test all generator diagrams and all those obtained from them by type A flypes with the freedom to apply type B flypes on each diagram before calculating ind_b.

To test some of the special generators with the most crossings required up to about half a day of calculation, and this made the work painful. In practice, it thus turned out necessary to speed up the procedure.

Some heuristical ways were used to more efficiently choose the sequence of vertices v and edges e. Briefly speaking, contracting edges far away from each other commutes, so we need to try only one order of contractions. Accounting for such redundancies made the calculation considerably faster at some places. It still took about 2 weeks to verify (8.1) for all generators and thus to complete the work for the genus 4 knots.

To give an impression why this simplification is necessary, we mention that much later we tried to recompute the result, for verification purposes, using the definition of ind_b as given here. This confirmed the outcome but took $3^1/_2$ months after splitting the generator list into 100 equal parts and processing them simultaneously. For some generators, the calculation took several weeks.

The knots up to 18 crossings served again as a test case and were checked several times with the different index implementations for runtime performance purposes. This concludes the proof of Theorem 8.1.1. □

Remark 8.3.3 There is a question of Rudolph whether a strongly quasipositive knot (or link) always has a strongly quasipositive (Bennequin) surface on the minimal number of strings. While we expect this not to be the case in general, we know of no counterexamples. The theorem shows that the answer is positive for (special) alternating knots of genus up to 4 or up to 18 crossings.

In an attempt at clarification, we summarize the 3 types of index (of *bipartite* graphs or diagrams) that occurred above:

$$\begin{array}{ccccc} & & \mathrm{ind}_b(D) & & \\ & & \wedge\vert & & \\ s(D)-\mathrm{mwf}(D) & \overset{!}{\geq} & \mathrm{ind}_0(D) & = & \mathrm{ind}(D)\,. \end{array} \tag{8.4}$$

The indices of the first, respectively second, line take, respectively do not take, into account keeping the bands. Thus, these indices can, respectively cannot, be used to estimate Bennequin surface string numbers. The indices of the left, respectively right, side take, respectively do not take, into account the distinction of vertices on the opposite side to an edge. The three indices are as argued additive under join (Theorem 2.9.4, Lemmas 7.3.6 and 8.2.4). The inequality marked with a "!" is known to become strict (i.e., not an equality) for some diagrams/graphs. The equality on the right uses Traczyk's aforementioned argument [Tr2], which we did not present here.

In [St13], we proved the existence of the minimal string Bennequin surface for alternating links of braid index 4. Note that our knot examples in [HS] have braid index 4, crossing number 16 and genus 3. So, if we exclude alternation, they would fall into *any* of the three categories on which we confirmed the Bennequin surface. Thus alternation is a crucial assumption. However, the proof of Theorem 8.1.1 and the Murasugi–Przytycki examples [MP, §19] of unsharp MWF inequality ($5 = \mathrm{mwf} < b = 6$) show the difficulties with the extension of the Bennequin surface result for alternating knots both of higher genus and braid index (> 4). Still, even if our construction of the Bennequin surface would fail giving minimum number of strings on alternating diagrams (of which we have no example yet), this would by far not mean that there is no minimal string surface. So, with some courage, we could ask:

Question 8.3.4 Does every alternating knot have a minimal string Bennequin surface?

Note that our tests admit more general application than our studied cases. For example, the sharp MWF and minimal string Bennequin surface criteria apply also for an alternating pretzel link $(x_1,\dots,x_n)$ with x_i odd and the twists therein reverse.

Dealing with canonical surfaces leads to some related question: Does every knot whose genus equals the canonical genus have a minimal string Bennequin surface? It is interesting to remark that for none of the examples reported in [HS] are these genera equal.

Chapter 9

The Alexander polynomial of alternating knots

While it is very well known what Alexander polynomials occur for an arbitrary knot, the question about those occurring for an alternating knot is much harder. We will apply our work to this problem in this chapter.

9.1 Hoste's conjecture

There is a conjecture I learned from personal communication with Murasugi, who attributes it to Hoste.

Conjecture 9.1.1 (Hoste) If $z \in \mathbb{C}$ is a root of the Alexander polynomial of an alternating knot, then $\Re e\, z > -1$.

Here we will use again our generator classification and an appropriate calculation to show the following.

Theorem 9.1.2 Hoste's conjecture is true for knots up to genus 4 (or equivalently for Alexander polynomials of maximal degree up to 4).

The parenthetic rephrasing refers to the normalization of Δ with $\Delta(t) = \Delta(1/t)$ and $\Delta(1) = 1$. (Beware that we will change this normalization during the proof.)

Some properties of the Alexander polynomials of alternating knots are known; see [Cw, Mu2, MS]. In particular, we note Theorem 2.8.4 regarding the coefficients.

The zeros seem less easy to control than the coefficients, and Hoste's conjecture appears open even for example for 2-bridge knots. Still, Theorem 2.8.4 will be a useful ingredient in the proof of Theorem 9.1.2. Also, there is one important tool to control some zeros of Δ, the signature.

Theorem 9.1.3 For any knot K the number of zeros of Δ_K in $\{z\in\mathbb{C}: |z|=1, \Im m z>0\}$, counted with multiplicity, is at least $\frac{1}{2}|\sigma(K)|$.

This is a folklore fact; see [Gr, St21] for some explanation. In [St16], it was proved alternatively for special alternating knots, and we deduced:

Corollary 9.1.4 ([St16]) If K is special alternating, then all zeros of Δ lie on $S^1=\{z\in\mathbb{C}: |z|=1\}$, and so Hoste's conjecture holds for K.

(The value $\Delta(-1)$ is the determinant, and non-zero for example by Theorem 2.8.4; but in fact it is never zero for any knot.)

One further important tool, Rouché's Theorem, will be introduced during the proof.

Proof of Theorem 9.1.2. Assume $\Delta(z)=0$ for $\Re e\, z\le -1$. Because by Theorem 2.8.4 no zeros of Δ lie on the negative real line, and furthermore Δ is real and reciprocal, we see that Δ must have the 4 distinct zeros $z^{\pm1}, \bar{z}^{\pm1}$. Then by Theorem 9.1.3, we see that Hoste's conjecture is true whenever $|\sigma|\ge 2g-2$, and in particular the case $g=1$ is finished.

Consider next $g=2$, $\sigma=0$. Let $\Delta_{[i]}=[\Delta]_{\min\deg\Delta+i}$. Then we have

$$\sum_{\Delta(z)=0} z = -\frac{\Delta_{[1]}}{\Delta_{[0]}} \ge 0. \tag{9.1}$$

However, if $\Re e\, z\le -1$, then the same holds for $\bar{z}$, and also $\Re e\, z^{-1}, \Re e\, \bar{z}^{-1}$ are negative, so that the left sum in (9.1) is in fact <-2, a contradiction. This finishes $g=2$.

The argument using (9.1) works also for $g=3$, $\sigma=2$ because the extra pair of zeros on S^1 augments the sum on the left of (9.1) by at most two, and so it is still negative.

For $g=3$ it remains to consider $\sigma=0$, the first non-trivial case.

Note that σ is constant on each series of alternating diagrams (as can be inferred from the combinatorial formula (2.41) for σ). Thus, we need to consider the series of $\sigma=0$ generators. Composite knots (inductively) and mutations are immaterial from the point of view of Hoste's conjecture, so again it suffices to consider prime generators and one diagram per generator.

Among the 4017 generators of genus 3, only 210 have $\sigma=0$. Now note (e.g., using (9.1)) that if $\Delta(z)=0$, $\Re e\, z\le -1$, then the remaining two zeros of Δ, different from $z^{\pm1}$ and their conjugates, must be real positive (and mutually inverse). Let z' be the (real) zero >1.

Now for any diagram D in the series $\langle D'\rangle$ of D' (resuming the notation of Definition 2.11.7), the Alexander polynomial has the form

$$\Delta(D) = \sum_{i=1}^{n} a_i \Delta(D_i) \cdot (1-t)^{n_i}, \tag{9.2}$$

where $a_i \in \mathbb{N}$, and D_i are obtained by smoothing out one crossing in n_i different $\sim$-equivalence classes of D'. Hereby, the convention for Δ was *changed*, so that $\min\deg\Delta = 0$ and $\Delta_{[0]} = \Delta(0) > 0$. An explanation of formula (9.2) is given [SSW]. It was clarified that all terms have leading coefficients of equal sign (and then the same applies to all other coefficients using Theorem 2.8.4).

The polynomials

$$\Delta_i := \Delta(D_i) \cdot (1-t)^{n_i} \quad \text{and} \quad \tilde{\Delta}_i := \Delta(D_i) \tag{9.3}$$

will be the central object of attention (and calculation) from now on. Note also that

$$n_i = \max\deg\Delta(D') - \max\deg\Delta(D_i) = 2g - \max\deg\Delta(D_i) = 6 - \max\deg\tilde{\Delta}_i\,. \tag{9.4}$$

In particular, $\Delta_i = 0$ if $n_i > 2g(=6)$, and these D_i and Δ_i can be discarded in the sum of (9.2). (We will soon see that it is very helpful to choose the number n of terms in this sum as small as possible.)

With this reduction, it is easy to see that

$$\begin{matrix} m_1 \\ M_1 \end{matrix} := \begin{matrix} \inf \\ \sup \end{matrix} \left\{ \left| \frac{\Delta_{[1]}(D)}{\Delta_{[0]}(D)} \right| : D \in \langle D' \rangle \right\} = \begin{matrix} \min \\ \max \\ i \end{matrix} \gamma(D_i), \tag{9.5}$$

where

$$\gamma(D_i) := \frac{(\Delta_i)_{[1]}}{(\Delta_i)_{[0]}} = \frac{(\tilde{\Delta}_i)_{[1]}}{(\tilde{\Delta}_i)_{[0]}} + 6 - \max\deg\tilde{\Delta}_i\,. \tag{9.6}$$

Then (9.1) implies that if there is a z of $\Re e\, z \le -1$ with $\Delta_D(z) = 0$, then the real zero $z' > 1$ of Δ_D satisfies

$$z' + \frac{1}{z'} > 2 + m_1\,. \tag{9.7}$$

Because z' is the unique zero > 1 (and because $2 + m_1 \ge 2$), (9.7) is equivalent to

$$\Delta(D)(z_0) < 0, \quad \text{where} \quad z_0 + \frac{1}{z_0} = 2 + m_1 \text{ and } z_0 > 1\,. \tag{9.8}$$

(Keep in mind that we normalized Δ so that $\Delta(0) > 0$, so the leading coefficient is also positive.) Now if we check that contrarily

$$\Delta(D_i)(z_0) > 0 \tag{9.9}$$

for all $i = 1, \ldots, n$, then (9.8) clearly cannot hold. Let us call (9.9) the *positive zero test*.

For each generator D' the D_i can be generated, m_1 and z_0 calculated and then (9.9) checked. (We excluded $n_i \geq 7$, and if $n_i = 6$ the only polynomial coming in question is $\tilde{\Delta}_i = 1$, so we can assume $n_i \leq 5$.)

The positive zero test took a few minutes and succeeded on the 210 generators. With this $g = 3$ is also finished.

For $g = 4$ more work is needed. As before $\sigma \geq 6$ is trivial, but for $\sigma = 4$ we have something to check. Now with two pairs of (conjugate) zeros of Δ on S^1, (9.1) and (9.5) give the condition

$$\left| \frac{\Delta_{[1]}(D)}{\Delta_{[0]}(D)} \right| < 2. \tag{9.10}$$

To see that this is false, by Theorem 2.8.4 and (9.4) it is enough to consider for each generator diagram D' the diagrams D_i with $n_i \leq 1$, and calculate that $\gamma(D_i) \geq 2$ for $\gamma(D_i)$ in (9.6). There are about 500,000 generators of $g = 4$ and $\sigma = 4$, and this test took a few hours, but succeeded.

For $g = 4$, $\sigma = 2$ we have a modification of the positive zero test used for $g = 3$, $\sigma = 0$. Most thoughts apply still, except that in (9.4) and (9.6) the "6" (standing for $2g$) becomes "8", and in (9.7) and (9.8) the additive 2 on the right disappears because again we have a new pair of zeros z on S^1, giving an additional contribution $2\Re e\, z$ up to 2 in the sum of (9.1). (Now we must also take care that $m_1 \geq 2$, so that z_0 in (9.8) is still real.)

The positive zero test thus now weakens (i.e., the condition tested is stronger and may more likely not hold), and there are more generators (about 190,000). There turn up to be 1157 cases of failure. In an attempt to handle these generators D', we enhanced the positive zero test.

Apply the test on the diagrams obtained from D' by $\bar{t}_2'$ twisting once (separately) at each $\sim$-equivalence class of D'. If the test succeeds on D_i', obtained from D' by a $\bar{t}_2'$ move at the i-th class, then we can exclude twists in D' at this class. With restrictions thus obtained we return again to the positive zero test of D'. The fewer classes to twist at mean that the set of diagrams D_i in (9.2) becomes smaller. Then m_1 in (9.5), and with also z_0 in (9.9), goes up while there are fewer positivity tests to perform.

With this trick from the 1157 cases, the number of difficult ones was reduced to 304. These, as well as the $\sigma = 0$ knots, require a new method, which we sought for a while and found in Rouché's Theorem.

Theorem 9.1.5 (Rouché's Theorem; see e.g., [RS, Chapter 1]) If two holomorphic functions f, g inside and on some piecewise smooth closed contour $C \subset \mathbb{C}$ satisfy

$$0 < |g(z)| < |f(z)| \quad \text{for all } z \in C, \tag{9.11}$$

then f and $f + g$ have the same number of zeros (with multiplicity) inside C.

For us f, g will be always sums of the type (9.2). We also would like to choose

$$C = \{\Re e\, z = -1\}. \tag{9.12}$$

Now, the contour is not closed. So let us argue why this choice is admissible.

In our case f,g will be polynomials of the same degree with real leading coefficients $\max\operatorname{cf} f$ and $\max\operatorname{cf} g$, which do not cancel out. Moreover, all ratios between coefficients are bounded, in a way depending only on the generator D'. Thus, it is not hard to a find a constant R (depending only on D'), such that the norms of f,g compare uniformly on $\{\Re e\, z \le -1, |z| = R\}$ in the same way as their absolute leading coefficients $|\max\operatorname{cf} f|$ and $|\max\operatorname{cf} g|$ compare. If $\max\operatorname{cf} f \ne \max\operatorname{cf} g$, we can achieve (9.11) after possibly swapping f,g. If $\max\operatorname{cf} f = \max\operatorname{cf} g$, we take g in (9.11) to be our present $g/2$ and apply the theorem twice.

With this understanding, we assume that C is chosen as in (9.12).

We will consider the property that

$$\Re e\,(f\bar{g}) \text{ is positive on all of } C. \tag{9.13}$$

This positivity condition is insensitive for the tricky case where we decided to divide g by 2. If (9.13) is satisfied, Rouché's Theorem implies that $f+g$ has the same number of roots z with $\Re e\, z \le -1$ as *either* of f or g do. For f this is precisely Theorem 9.1.5, assuming with the preceding argument that (9.11) holds. For g one applies that theorem on $f+g$ and $-g$ and uses (9.13) to ensure (9.11). (Clearly, $z \in C$ is not a problem in either case because of (9.11).)

Therefore, to confirm Hoste's conjecture on $\langle D' \rangle$, it suffices that (1) the constant polynomial 1 is among the $\tilde{\Delta}_i$; and that (2) all conditions

$$\Re e\left(\Delta_i(z)\Delta_j(\bar{z})\right) > 0 \quad \text{for } z \in C \text{ and all } 1 \le i < j \le n,$$

are satisfied with Δ_i as in (9.3). Then we can conclude the property that the number of roots z left of C is 0 by induction on, say, $\Delta_{[0]}$. We call the above two conditions in the following the *Rouché test*.

Now the number of checks grows quadratically in n, which makes the reduction of the number of Δ_i much more urgent than before. We already discarded zero polynomials. Clearly, duplicate Δ_i, even up to scalars, can be discarded too, and one easily sees that this is true also up to multiplication by powers of $1-t$. So we divided all factors $1-t$ out of the Δ_i prior to discarding duplicates. Because the origin of the values n_i in (9.3) will be irrelevant for the rest of the calculation, assume w.l.o.g. that $\tilde{\Delta}_i$ are the polynomials reduced in the aforementioned way (indivisible by $t-1$). By symmetry one sees then that $\max\deg\tilde{\Delta}_i$ is an even number (between 0 and 8).

For the purpose of testing the 304 remaining $\sigma = 2$ generators, we used Mathematica™ [Wo] and the Rouché test with the so far reduced sets of $\tilde{\Delta}_i$ for each D'. In determining the $\tilde{\Delta}_i$, we also used the restrictions obtained in the enhanced positive zero test on which $\sim$-equivalence classes of the generator twists are necessary. After an hour and 15 minutes, Mathematica reported success on all 304 generators D', thus completing the $\sigma = 2$ case.

For $\sigma = 0$, however, there are about 60,000 generators (with no restrictions on $\sim$-equivalence classes to twist at), and Mathematica, albeit very intelligent, is known to pay for its intelligence with its speed. This made further reductions of the set of $\tilde{\Delta}_i$ necessary.

In this vein, note that if some $\tilde{\Delta}_i$ lies in the convex hull of others, it is also redundant. (It is enough that a_i remain positive in (9.2); integrality is inessential.) Unfortunately, the effort of identifying exactly convex linear combinations as a pre-reduction to the Rouché test would take more effort than the test itself.

In practice, we did the following for polynomials $\tilde{\Delta}_i$ of degrees $d = 2,4$. We determined the minimal and maximal ratios of $(\tilde{\Delta}_i)_{[j]}/(\tilde{\Delta}_i)_{[0]}$ for $1 \le j \le d/2$. Then we replaced the set of all $\tilde{\Delta}_i$ of degree d by a collection of $2^{d/2}$ polynomials given by taking for each coefficient $\Delta_{[j]}$ with $1 \le j \le d/2$ once the minimal, and once the maximal ratio, setting $\Delta_{[0]} = 1$ (and completing the other coefficients by symmetry $\Delta_{[d-j]} = \Delta_{[j]}$).

For $d = 2$ this replacement is equivalent (under convex linear combinations), so it is a genuine simplification, but this did not speed up the check sufficiently. Contrarily, for $d = 4$ the new polynomials have a larger convex hull, which augments the risk of failure of the test. However, using only the 4 polynomials instead of the dozens of others made the test about 10 times faster. It still took a couple of days to carry out.

Mathematica reported success of the last, fastest, form of the test on all $\sigma = 0$ generators, except only one, the knot 12_{1039} of [HT]. There the test worked out at least using the full list of polynomials $\tilde{\Delta}_i$ of degree 4 rather than their 4 substitutes.

With this, after a few weeks of work, the check of Hoste's conjecture in genus ≤ 4 was completed. □

Remark 9.1.6 After we finished our proof, we learned of Ozsváth and Szabó's inequality (9.17). It would replace the r.h.s. of (9.1) by 2, thereby settling $g = 3$, $\sigma = 0$ directly, and sharpening the positive zero test for $g = 4$. Also, (9.10) would be excluded straightforwardly. Still, both the nature of the tests and the extent of calculation would not be reduced significantly.

Note that (9.5) can be extended easily to $\left|\Delta_{[i]}/\Delta_{[0]}\right|$ for $i > 1$. In particular we have this property:

Proposition 9.1.7 For given $i \ge 1$ and $g > 0$, the ratios $\left\{\, \left|\Delta_{[i]}/\Delta_{[0]}\right| \,:\, g(K) = g \,\right\}$ are bounded. □

Such a property is not at all evident without using generators. Of course, it is possible, using, e.g., Corollary 3.4.2, to give explicit estimates in terms of g. Moreover, we gain a practical method to calculate (even sharp) upper and lower bounds on this ratio for a given genus when the generators are available.

Example 9.1.8 For example, for $g = 4$, $\sigma = 0$ prime alternating knots we found

$$3 \le -\frac{\Delta_{[1]}}{\Delta_{[0]}} \le 20, \quad 5 \le \frac{\Delta_{[2]}}{\Delta_{[0]}} \le 122, \quad 7 \le -\frac{\Delta_{[3]}}{\Delta_{[0]}} \le 333, \quad \text{and} \quad 8 \le \frac{\Delta_{[4]}}{\Delta_{[0]}} \le 461,$$

and these bounds are the best possible (except I did not check if equalities are attained, or the inequalities are always strict). They were found by unifying the analogous bounds that can be obtained for each generator separately. Using the latter, we also estimated for each generator the Jensen integral for the Euclidean Mahler measure $M(\Delta)$, the product of norms of zeros of Δ outside the unit circle (see [SSW]). In combination, we have

$$M(\Delta) \leq 638.21$$

for the Euclidean Mahler measure of $g = 4$, $\sigma = 0$ prime alternating knots. (This estimate may not be very sharp.)

Such data were obtained in an earlier attempt at this case of Hoste's conjecture, prior to using Rouché's theorem. They can be extended with the proper (not small, but manageable) amount of calculation to all $g = 4$ alternating knots. Note again that the special alternating knots (here $\sigma = 8$) are mostly clear from this point of view. Even for arbitrary genus g, it is easy to deduce from the preceding discussion that

$$0 \leq (-1)^j \frac{\Delta_{[j]}}{\Delta_{[0]}} < \binom{2g}{j},$$

with the right inequality the best possible and the left bound not improvable beyond 1. (The bound is exactly 1 for $j = 1$ by the work in [MS]; apparently this is true also for higher j and would follow from Fox's Trapezoidal Conjecture as discussed in [HS, St16] and right below.) Also $M(\Delta) \equiv 1$.

By modifying the contour C in Rouché's test one could obtain further more specific information about location of Δ-zeros on certain families of knots.

9.2 The log-concavity conjecture

Call a polynomial $X \in \mathbb{Z}[t^{\pm 1}]$ to be log-*concave* if its coefficients $[X]_k$ are log-concave, i.e.,

$$[X]_k^2 \geq [X]_{k+1}[X]_{k-1} \tag{9.14}$$

for all $k \in \mathbb{Z}$. The last open problem we consider is the log-concavity conjecture. This conjecture, made in [St16], states:

Conjecture 9.2.1 Let $n(K)$ be the number of components of a link K.

(1) If K is an alternating link, then $t^{(1-n(K))/2}\Delta_K(t)$ is log-concave.

(2) If K is a positive link, then $t^{(1-n(K))/2}\nabla_K(\sqrt{t})$ is log-concave.

We will refer to both properties as "Δ-log-concavity", respectively "∇-log-concavity."

We recall that by Crowell–Murasugi [Cw, Mu2] for Part 1, when $\Delta_K(t)$ is normalized so that $\min\deg\Delta_K = 0$ and $[\Delta_K]_0 > 0$, the polynomial $\Delta_K(-t)$ is positive, i.e., all its coefficients are non-negative. The same property holds by Cromwell [Cr] for Part 2 and $t^{(1-n(K))/2}\nabla_K(\sqrt{t})$.

Part 1 is a natural strengthening of *Fox's Trapezoidal conjecture.*

Conjecture 9.2.2 (Fox) If K is an alternating knot, then there is a number $0 \le n \le g(K)$ such that we have

$$\begin{aligned} \left|\Delta_{[k]}\right| &> \left|\Delta_{[k-1]}\right| \quad \text{for } k = 1,\dots,n, \\ \left|\Delta_{[k]}\right| &= \left|\Delta_{[k-1]}\right| \quad \text{for } k = n+1,\dots,g(K). \end{aligned} \tag{9.15}$$

We call polynomials of this form *trapezoidal.* A similar property can be conjectured for (non-split) alternating links, replacing $g(K)$ by $\lfloor \operatorname{span}\Delta_K/2 \rfloor$.

The Trapezoidal conjecture has received some treatment in the literature, being verified for 2-bridge knots [Ha] (see also [Bu]) and later for a larger class of alternating algebraic knots [Mu6]. More recently, Ozsváth and Szabó used their knot Floer homology [OS] to derive a family of linear inequalities on the coefficients of Δ for an alternating knot. (Keep in mind (2.39).)

Proposition 9.2.3 (Ozsváth and Szabó) Let K be an alternating knot of signature $\sigma = \sigma(K)$ and genus $g = g(K)$, and let $\Delta = \Delta(K)$ be normalized so that $\min\deg\Delta = 0$ and $[\Delta]_0 > 0$. Then for each integer $s \ge 0$,

$$(-1)^{s+g} \sum_{j=1}^{g-s} j \cdot [\Delta]_{s+g+j} \le (-1)^{s+\sigma/2} \max\left(0, \left\lceil \frac{|\sigma| - 2s}{4} \right\rceil\right). \tag{9.16}$$

For genus $g = 2$ the inequalities (9.16) are very similar to (and slightly stronger than) the Trapezoidal conjecture, and for $s = g - 2$ yield

$$-[\Delta]_1 \ge 2[\Delta]_0 + \begin{cases} -1 & \text{if } |\sigma| = 2g \\ +1 & \text{if } |\sigma| = 2g-2 \\ 0 & \text{otherwise} \end{cases}, \tag{9.17}$$

which settles in (9.15) the case $k = 1$ (for knots):

$$\Delta_{[0]} \le -\Delta_{[1]}. \tag{9.18}$$

More precisely we have the following:

Proposition 9.2.4 For every alternating knot K, (9.18) holds with equality if and only if K is the $(2, 2g+1)$-torus knot.

Proof. By (9.17) we have for $-\Delta_{[1]} = \Delta_{[0]}$ that $\sigma = 2g$, hence K is special alternating (by Proposition 2.10.1), and

$$\Delta_{[0]} = -\Delta_{[1]} = 1. \tag{9.19}$$

Now by Corollary 9.1.4 (as discussed in [St16]) Δ_K is log-concave, and with (9.19) all $\Delta_{[i]} = \pm 1$. Because Δ is also alternating (by Theorem 2.8.4), we have

$$\left|\Delta_K(-1)\right| = 2g(K) + 1 \le c(K).$$

By a well-known result, first due to Crowell [Cw2], the only alternating knots K with such polynomial are the $(2,n)$-torus knots K (with $n = c(K) = 2g+1$). □

For $g > 2$ and $s < g-2$, the inequalities (9.16) do not relate directly to the Trapezoidal conjecture. In-Dae Jong [Jn] has proved the Trapezoidal conjecture independently from us up to genus 2 using the generator description in Theorem 2.11.11, and it was observed that for genus 2 the log-concavity of Δ easily follows from trapezoidality.

In [St16], we showed Part 2 of Conjecture 9.2.1 for special alternating knots by giving a new proof of Theorem 9.1.3 in this case.

We have now the following.

Theorem 9.2.5 Both Part 1 and 2 of Conjecture 9.2.1 (and, therefore, also Fox's conjecture) hold for knots of genus at most 4.

The preceding work easily settles the positive case.

Proof of Part 2 of Theorem 9.2.5. From Theorem 9.1.3 one easily sees that all zeros of $\nabla(\sqrt{t})$ are real if $\sigma(K) \ge 2g(K) - 2$. With the explanation in [St16], we then have log-concavity of ∇. Thus, we need to check just the polynomial of 14_{45657} directly. □

Part 1 is much harder to check because we have no easy sufficient conditions at hand. We explain how we proceeded.

Proof of Part 1 of Theorem 9.2.5. We had to test all generators up to genus 4. A small defect of log-concavity is that it is not straightforwardly preserved under product (as is trapezoidality), so composite generators must be handled as well.

Due to its volume, the check had to be optimized strongly. Consider a particular generator diagram D. We use the notation of (9.3) in the proof of Theorem 9.1.2. Because one can recover Δ_i by

$$\Delta_i = (1-t)^{g(D) - \max\deg\tilde{\Delta}_i/2}\tilde{\Delta}_i \tag{9.20}$$

even when $\tilde{\Delta}_i$ is taken up to powers of $1-t$, we can assume w.l.o.g. that we divide out all factors $1-t$ in $\tilde{\Delta}_i$ so that $\tilde{\Delta}_i(1) \ne 0$ and $[\tilde{\Delta}_i]_0 > 0$. First, we designed the calculation of $\tilde{\Delta}_i$ so that diagrams D_i of genus 0 or split ones (where $\tilde{\Delta}_i$ is zero or a scalar) are detected and discarded in advance. For the calculated $\tilde{\Delta}_i$, those of small degree are again treated extra. For degree 0, it is enough to keep the polynomial 1. For degree 2, we need only two polynomials, those whose ratio $[\tilde{\Delta}_i]_1/[\tilde{\Delta}_i]_0$ is minimal and maximal.

The $\tilde{\Delta}_i$ of degree 4 lead to the main reduction. In this case, we have a number of points

$$\big([\tilde{\Delta}_i]_1/[\tilde{\Delta}_i]_0,\,[\tilde{\Delta}_i]_2/[\tilde{\Delta}_i]_0\big)\in\mathbb{R}^2 \tag{9.21}$$

in the plane. We need to keep only the *extremal* ones, i.e., those which span the convex hull, and discard the others (which lie in the convex hull). In practice, it turned out that this reduced, for a fixed generator, the number of $\tilde{\Delta}_i$ of degree 4 up to a factor of 150 and the number of all $\tilde{\Delta}_i$ up to a factor of 11.5.

With the so reduced set of $\tilde{\Delta}_i$, we recover Δ_i by (9.20) and must test for $1\le k\le g(D)$

$$[\Delta_i]_k^2\ge[\Delta_i]_{k-1}[\Delta_i]_{k+1}\,, \tag{9.22}$$

and the set of such conditions for convex linear combinations of Δ_i and $k<g(D)$. (The case $k=g(D)$ for linear combinations follows directly from (9.22) and symmetry of Δ_i.) The latter again turns into deciding whether the convex polygon $\Sigma\subset\mathbb{R}^2$ spanned by points

$$\big(-[\Delta_i]_{k+1}/[\Delta_i]_k,\,[\Delta_i]_{k+2}/[\Delta_i]_k\big)\in\mathbb{R}^2$$

is contained in the region $R=\{\,(x,y)\,:\,x\ge 0,\,y\le x^2\,\}$. Clearly, it is enough to test the inclusion $\partial\Sigma\subset R$ for the boundary $\partial\Sigma$ of Σ. This boundary is a certain collection of edges between extremal points and can be determined similarly to the convex hull reduction of the $\tilde{\Delta}_i$ of degree 4. Finally, we have to check that the function $\mathbb{R}^2\ni(x,y)\mapsto x^2-y$ is non-negative on these edges. In practice, we made the tests by an ε stricter to avoid floating point number deviations.

For a handful of pairs (D,k), we have $\Sigma\not\subset R$. Let us call such pairs and the corresponding generators D *irregular*. For genus 3, the irregular generators are 7_1 and 8_5. They still do not mean that we have a counterexample because the points coming from Δ polynomials of diagrams in the series of D form only a subset of Σ. In such cases, we again check log-concavity of the generator polynomial $\Delta(D)$, and then apply the described test to diagrams D' obtained by one $\bar{t}'_2$ twist from D. Iterating this a few times shows that only one explicit one-parameter family, of $(3,3,2m)$ pretzel diagrams ($m>0$), remains, which can be verified directly.

Among prime generators of genus 4, only 3 are irregular, and they are 9_1, 10_2 and 10_{46}. Using the twisted diagram check, they leave out only one one-parameter family each, consisting of the rational knots $2m-1,8$ and the pretzel knots $P(2m,1,7)$ and $P(2m,3,5)$ (with $m>0$) respectively. Again, the latter are easily settled directly. Among composite generators up to genus 4, no irregular ones occur.

Despite its theoretical simplicity, the check involved much skillful implementational effort. Among others, we decided to write an own (optimized) procedure for calculating the Alexander polynomial, instead of using the functionality of [HT], which applies a substitution to the skein polynomial obtained by the Millett–Ewing algorithm.

With all optimization, we had a speed-up by a factor of almost 5 in comparison to the initial implementation. Finally, while genus 3 could be done in about 8 minutes, we could reduce the effort for genus 4 only to about a second for the simplest generators, and up to 30 seconds for the most complicated ones. This meant that the verification of the full list of generators had to be done in a few dozen parts with a couple of them running over several weeks. □

Remark 9.2.6 Note that for Fox's conjecture the test could be simplified because dealing with linear combinations of the Δ_i becomes obsolete. (The difficulty of log-concavity is that (9.14) is not a linear condition.) Still this would not have reduced the magnitude of calculation significantly because the most time is required for the calculation of the Δ_i.

With some effort to adapt the computation to links, we obtained the following outcome.

Theorem 9.2.7 The Δ-log-concavity (and hence Fox's Conjecture 9.2.2) holds for n-component alternating links L with $\operatorname{span}\Delta(L)+n \leq 9$ (or equivalently $g(L)+n \leq 5$).

Proof. We recurred the calculation from generators of these links to knot generators by using claspings (3.3). The check turned out far more laborious than for knots, but there is no real difference in the method, so we skip details to save space. □

Remark 9.2.8 It is suspectable that in Conjecture 9.2.2 we always have for a knot K

$$n \geq g(K) - |\sigma(K)/2|\,. \tag{9.23}$$

It is easily observed that it suffices to check (9.23) for the generators and that prime ones are enough. Moreover, the inequality is trivial for special alternating generators, so consider only the non-special ones. We tested that all such generators up to genus 4 satisfy (9.23). (For $g=2$, and also $n=0$, this property is implied by Ozsváth–Szabó's inequality (9.17).) Actually, very few generators have $n < g(K)$: Only 4 of the non-special prime generators of genus 3 and 30 of those of genus 4 have coefficients showing a true "trapez" rather than a "triangle." We may mention that we did not test any analogue of (9.23) for links.

Remark 9.2.9 A minor addition to Part 1 of the log-concavity Conjecture 9.2.1 is that if we have equality in (9.14), then all three of $[X]_k$, $-[X]_{k\pm 1}$ are equal. Our test verified this extra property, too, for knots up to genus 4 and the links in Theorem 9.2.7.

9.3 Complete linear relations by degree

9.3.1 Linear relations up to constants

Let us remark that our proof of the Fox conjecture can be conceptually extended. While it is clear that one can verify a particular sort of linear inequality between coefficients of Δ for a given genus, we can in fact determine the *complete list* of such inequalities.

Definition 9.3.1 A *(convex) polytope* Σ is the convex hull of a finite number of points in $\mathbb{R}^n$ and *rays* $\mathbb{R}_+ \cdot \vec{b}$ for $\vec{b} \in \mathbb{R}^n$. Alternatively, one may describe a polytope by a set of linear inequalities (or as intersection of half-spaces of $\mathbb{R}^n$). If no rays occur, we call Σ *bounded.* If at most one point is there (with no point meaning that the vertex is the origin), then Σ is a *cone.* A *facet* of Σ is a codimension-1 subset of the boundary $\partial\Sigma$ of Σ given by intersection of Σ with a hyperplane.

There seems no uniform rule for the usage of "polytope" and "polyhedron" in the literature. Our attitude here is to avoid the latter term throughout. The word "convex" will be usually omitted because all polytopes we will deal with are such. Also note that, in contrast to us, many authors consider a polytope to be bounded per definition. Our terminology needs one more important clarification.

Convention. In the following the term *convex hull* $\operatorname{conv}(D)$ of a set $D \subset \mathbb{R}^n$ is meant as *the closure* of the set

$$\Big\{ \sum_{i=1}^{m} \lambda_i x_i \,:\, \sum_{i=1}^{m} \lambda_i = 1,\, \lambda_i \geq 0,\, x_i \in D \Big\}.$$

The reason for taking closures is easily seen from the example of a line and a point (outside the line) in the plane. This phenomenon wildens in higher dimension: For a non-closed convex polytope, only parts of some facets would be there. Formulated in terms of the linear inequalities describing the polytope, certain points for which some inequality is exact (i.e., an equality) are allowed while others are not. The discussion of these cases puts no reasonable use-over-effort ratio in prospect. So we assume *all convex sets are closed* (and the inequalities describing the polytope are always non-strict).

We shall treat here genus 3 exemplarily. For technical reasons we formulate the result in terms of ∇.

Proposition 9.3.2 The signed coefficients $\nabla_k = [\nabla]_k \cdot \operatorname{sgn}([\nabla]_6)$ of $\nabla(K)$ for an alternating knot K of genus 3 satisfy the following 6 inequalities:

$$\begin{array}{rclrcl} \nabla_4 - \dfrac{\nabla_2 + 5\nabla_6}{2} & \leq & \dfrac{1}{2}, & \nabla_4 - \nabla_2 - 3\nabla_6 & \leq & 2, \\ \nabla_4 + 3\nabla_2 + 9\nabla_6 & \geq & -4, & \nabla_4 + \dfrac{\nabla_2}{2} + 4\nabla_6 & \geq & -\dfrac{1}{2}, \\ \nabla_4 - \nabla_2 + 7\nabla_6 & \geq & -7, & \nabla_2 - 3\nabla_6 & \leq & 4. \end{array} \qquad (9.24)$$

This set of linear inequalities is complete up to constants, in the sense that any other linear inequality valid for all (or all but finitely many) alternating genus 3 knots is, up to a worse absolute term, a consequence of those above.

We will explain later how to remove the defect of the constants, but it requires a bit more sophisticated calculation.

Proof. If for each generator D we take all $\tilde{\Delta}_i \neq \Delta(D)$, then all have degree at most 4. (We will deal with the $\Delta(D)$ later.) Let $\nabla_i(z)$ be given by

$$\nabla_i(t^{1/2} - t^{-1/2}) = \Delta_i(t) \cdot t^{-\max\deg \Delta_i/2}.$$

The reason for using $\nabla_i(z)$ is that the factor $1-t$ in (9.20) turns into z, and so we have just a degree shift, which is faster on the computer. (Thus, in fact, we used ∇_i instead of Δ_i almost throughout the calculation for Theorem 9.2.5.) Now $\max\deg \tilde{\Delta}_i = \operatorname{span} \nabla_i$ (recall §2.8).

Now we can do the previous sort of determination of extremal polynomials ∇_i of span 4, but this time with all generators taken together. We again collect the corresponding (extremal) points in the plane

$$(x,y) = \big([\nabla_i]_2/[\nabla_i]_6,\, [\nabla_i]_4/[\nabla_i]_6\big) \in \mathbb{R}^2, \qquad (9.25)$$

as in (9.21), together with the two points (with $x = 0$) describing the convex hull (which now is an interval) for the ∇_i of span 0 and 2. Here is the result:

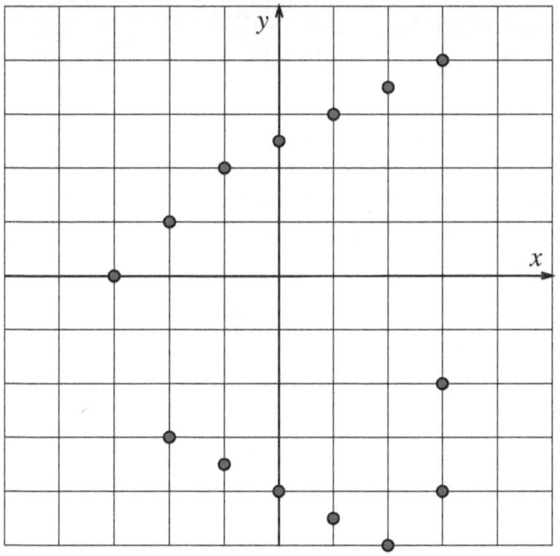

(9.26)

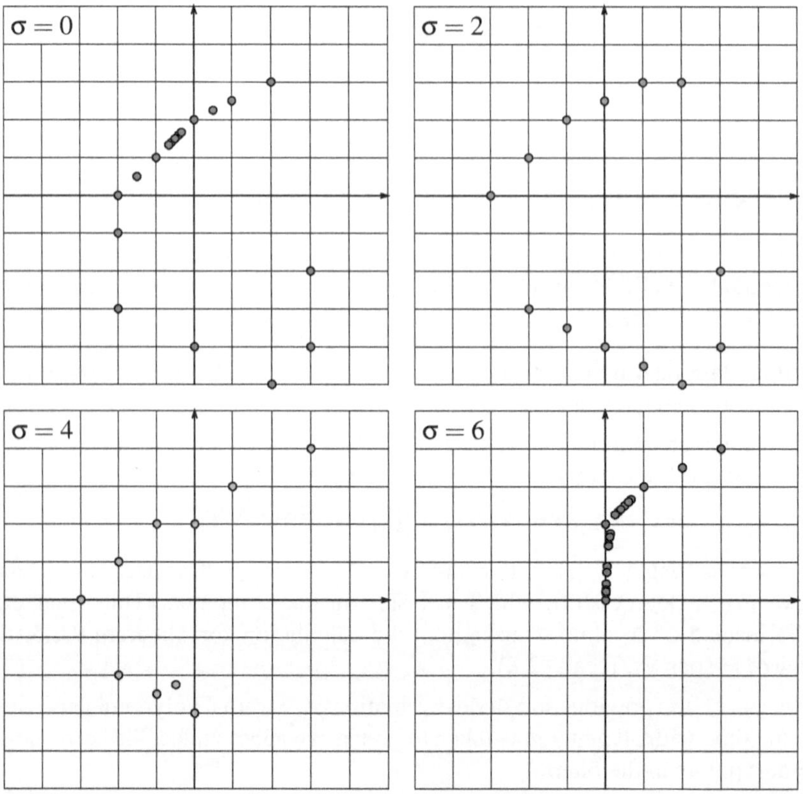

FIGURE 9.1: Let $\tilde{\Sigma}_D = \text{conv}\,\{(x,y)\}$ with (x,y) as given in (9.25) and below it, taken for a generator D. The charts show extremal points of the polygons $\tilde{\Sigma}_D$ for generators D of genus $g = 3$ and given signature σ. It is evident that a few extremal points of some $\tilde{\Sigma}_D$ are not extremal for $\tilde{\Sigma}_{g,\sigma} = \text{conv}\left(\bigcup_D \tilde{\Sigma}_D\right)$.

The convex hull $\tilde{\Sigma}$ is a hexagon, described by equations (one for each boundary edge) in which the r.h.s. of an equation in (9.24) is set to 0, ∇_2 is replaced by x, ∇_4 by y, and ∇_6 by 1. To obtain (9.24), we need then to calculate the minimal or maximal value of the l.h.s. over all generator polynomials $\Delta(D)$ (which we previously omitted from the calculation).

To see completeness, note that for each $\Delta_i \neq \Delta(D)$ one has a sequence of diagrams $\bar{D}_j \in \langle D \rangle$ with

$$\lim_{j\to\infty} \frac{\Delta(\bar{D}_j)}{[\Delta(\bar{D}_j)]_0} = \frac{\Delta_i}{[\Delta_i]_0},$$

so the set of points (9.25) is not reducible. □

9.3.2 General genus

Let us write more generally $\tilde{\Sigma}_g \subset \mathbb{R}_+^{g-1}$ for the convex sets obtained as in the preceding proof for generators of genus g. For $g > 3$ the procedure remains largely the same. Note that these polytopes will always be bounded due to Proposition 9.1.7.

It is natural to restrict oneself also to knots of given signature, and Figure 9.1 shows the pictures corresponding to (9.26) for $g = 3$. For example, for $\sigma = 6$ (the special alternating knots), one derives the (complete up to constants linear) inequalities:

$$\begin{aligned} \nabla_4 - \tfrac{4}{3}\nabla_2 &\geq -3\,, & \nabla_2 &\geq 3\,, \\ \nabla_4 - \nabla_2 - 2\nabla_6 &\leq -2\,, & \nabla_4 - \frac{\nabla_2 + 5\nabla_6}{2} &\leq -\frac{1}{2}\,. \end{aligned} \tag{9.27}$$

Let us denote the corresponding (bounded) polytopes as $\tilde{\Sigma}_{g,\sigma}$.

The graphics of Figure 9.1 were displayed to show the difference to the properties considered at venues unrelated to our approach. Roughly (i.e., ignoring the additive terms independent of ∇_k), the conditions (9.16) of Ozsváth–Szabó (only $s = 0, 1$ are relevant here) would lead to the restrictions

$$y \leq 4 \quad \text{and} \quad y \leq \frac{x+6}{2}\,, \tag{9.28}$$

while the trapezoidal inequalities (9.15) (with the options $>$ and $=$ merged into $\geq$) transcribe into

$$y \leq \min\left(5, \frac{x+5}{2}, \frac{x}{3} + 3\right). \tag{9.29}$$

These inequalities are (as should be) easily seen to follow from the ones describing our polygons $\tilde{\Sigma}$, but they are much weaker consequences. Note in particular that, even taken together, (9.28) and (9.29) restrict the possible (x,y) only to an infinite region of the plane. Also, the dependence of the shape of ∇ on the signature becomes much more apparent in our polygons than in (9.16). This shows how much further information one can obtain with our method if one considers a fixed genus. Still (9.15) or (9.16) may define (without abusing constants) a facet of Σ. We will test this explicitly below for $g = 3$.

9.3.3 Removing inaccuracy up to constants

If one wants to remove the inaccuracy up to constants, we have the following statement, in general dimension, which is worth taking record of.

Theorem 9.3.3 The (closed) convex hull Σ_g of the set of Conway polynomials of alternating knots of given genus g, or $\Sigma_{g,\sigma}$ given genus and signature, is a convex polytope in $\mathbb{R}^g$.

Proof. To determine Σ_g (and similar for $\Sigma_{g,\sigma}$), one must find the convex hull $\tilde{\Sigma}_D$ of the points (9.25) for each generator D separately, take the cone $\mathbb{R}_+ \cdot (\tilde{\Sigma}_D \times \{1\})$ of $\tilde{\Sigma}_D$ in $\mathbb{R}^g$ and translate this cone to have the summit $\nabla(D)$. Then Σ_g is the convex hull of the union of all these translated cones taken over all generators D (for $\Sigma_{g,\sigma}$ only those with with $\sigma(D) = \sigma$). This can be rewritten as

$$\Sigma_g = \mathbb{R}_+ \cdot (\tilde{\Sigma}_g \times \{1\}) + \mathrm{conv}\left(\{\nabla(D) : D \text{ is a generator of genus } g\}\right) \tag{9.30}$$

(with the $+$ being the Minkowski set-theoretic sum $A + B = \{a + b : a \in A, b \in B\}$). The result Σ_g is a(n unbounded) convex polytope as follows from the Farkas–Minkowski–Weyl Theorem for convex polytopes (see e.g., [Sc], Corollary 7.1a). □

Remark 9.3.4 Note that one can replace "alternating" by "positive" in Theorem 9.3.3. (For knots of given genus and signature a bit extra argument is needed.) We chose not to elaborate on the positive case here, though.

Let us call a facet $F \subset \partial\Sigma$ of $\Sigma \subset \mathbb{R}^g$ to be *n-open* if it contains an affine-translated copy of $\mathbb{R}_+^n \times \{0\}^{\times g-n}$ but not one of $\mathbb{R}_+^{n+1} \times \{0\}^{\times g-n-1}$. Say F is *largest* if it is $(g-1)$-open. Then inequalities like the above (9.24) or (9.27), said to be complete up to constants, are in fact those for the largest facets of Σ_g.

Theorem 9.3.3 clearly also gives a practical way to determine Σ_g, though it requires to work in one dimension up in comparison to $\tilde{\Sigma}_g$. The latter would likely be quite more complicated to describe precisely for high genus. (In calculating $\tilde{\Sigma}_3$ above, we took advantage of the 2-dimensional convex hull algorithm we had implemented for Theorem 9.2.5.) Nevertheless, this is feasible (even by hand) for genus 2 and was carried out by In-Dae Jong [Jn2]. The result can be stated as follows.

Theorem 9.3.5 The complete set of linear inequalities satisfied by the coefficients $\nabla_i = [\nabla]_i \cdot \mathrm{sgn}([\nabla]_4)$ of Conway polynomials of alternating genus 2 knots of given signature is:

$$\nabla_4 \geq 1 \text{ (for all } \sigma\text{) and } \left\{ \begin{array}{rl} -2\nabla_4 - 1 \leq \nabla_2 \leq \nabla_4 + 1 & \text{for } \sigma = 0, \\ -2\nabla_4 + 1 \leq \nabla_2 \leq 2\nabla_4 - 1 & \text{for } |\sigma| = 2, \\ 2 \leq \nabla_2 \leq 2\nabla_4 + 1 & \text{for } |\sigma| = 4 \end{array} \right\}.$$

For genus 3, one must use a computer. There is software for converting a convex-hull (vertex-ray) representation of a polytope like (9.30) into a half-space intersection (linear inequalities) representation. We used the programs `cdd/cdd+` of Komei Fukuda [Fu] and `lrs` of David Avis [Av] to find a linear inequality representation for $\Sigma_{3,\sigma}$ from (9.30) and the calculation of Figure 9.1. The programs do essentially the same but use different algorithms. We ran them both and compared the output for consistency security (except in the rather long case $g = 4, \sigma = 8$, where we used only `lrs`).

The description of $\Sigma_{3,\sigma}$ is shown in the following table in columns 2 and 3. An entry of the form *mvnr* in the third column means that $\Sigma_{3,\sigma}$ is the convex hull of m vertices and n rays. (For example, 1v2r is a planar angle segment, and 2v1r is a plane half-strip.) The second column gives the number of facets of (or minimal number of linear inequalities describing) $\Sigma_{3,\sigma}$.

The following columns show the efficiency of the conditions (9.16) (with OSn standing for the case $s = n - 1$) and (9.15) (with Tm standing for the inequality obtained by putting $k = m$ and joining the two alternatives into a non-strict inequality, without regard to the other k) for $g = 3$ and given signature σ. An entry of the form mvnr is as in column 2 and describes here the intersection of $\Sigma_{3,\sigma}$ with the hyperplane defined by the inequality being exact (i.e., an equality). In case the intersection is empty (the inequality is never exact), a bracketed number [n] means that the smallest defect of the absolute term is n. (That is, in an inequality $x \le y$ with $x \in \mathcal{L}in\{\nabla_i\}$ and $y \in \mathbb{R}$ the largest value of the l.h.s. on $\Sigma_{3,\sigma}$ is $y - n$.) One sees that both the trapezoidal and Ozsváth–Szabó inequalities are (not more and not less than) moderately good as conditions on the Alexander polynomial for genus 3, and their sharpness increases with increasing (absolute) signature.

σ \ ineq.	f	vr	OS1	OS2	T1	T2	T3
0	10	5v7r	[2]	[1]	[2]	[2]	[2]
2	12	6v8r	[1]	1v	[1]	[1]	2v2r
4	8	4v5r	[1]	2v1r	[2]	1v1r	2v2r
6	6	3v4r	2v	1v1r	1v	1v1r	2v2r
total	12	6v6r	–	–	1v	2v1r	5v2r

The determination of $\Sigma_{3,\sigma}$ also leads to a description of Σ_3, which is stated in the below theorem. Some data is given also in the bottom line of the table. Note that (9.16) depend, at least in the constant terms, slightly on σ. Thus it seems less helpful to compare Σ_3 to these inequalities directly. Moreover, keep in mind that also in Tn our stipulation $\nabla_6 > 0$ leads to a sign choice of ∇_i, which depends on σ mod 4 (as we worked out in §2.10; see (2.39)).

Theorem 9.3.6 The complete linear inequalities satisfied by the signed Conway coefficients $\nabla_k = [\nabla]_k \cdot \mathrm{sgn}([\nabla]_6)$ of an alternating knot of genus 3 are those in (9.24), together with the following 6 other inequalities:

$$\begin{aligned} 40\nabla_6 - \nabla_4 - 12\nabla_2 &\ge -37, & 34\nabla_6 + 8\nabla_4 + 3\nabla_2 &\ge -1, \\ 5\nabla_6 + \nabla_4 + \nabla_2 &\ge 0, & 3\nabla_6 + \nabla_2 &\ge -1, \\ 9\nabla_6 - 3\nabla_4 + \nabla_2 &\ge 0, & \nabla_6 &\ge 1. \qquad \square \end{aligned}$$

We conclude with the table for $g = 4$.

ineq. / σ	f	vr	OS1	OS2	OS3	T1	T2	T3	T4
0	105	26v26r	1v	[2]	1v	[1]	[1]	[1]	[1]
2	115	29v31r	[1]	[2]	[1]	[2]	[2]	[2]	4v4r
4	81	21v24r	1v	[1]	1v	[1]	[1]	2v2r	5v5r
6	41	11v14r	2v1r	[1]	2v1r	[2]	1v1r	2v2r	5v5r
8	18	5v8r	1v1r	3v	1v1r	1v	1v1r	2v2r	2v4r
total	100	27v26r	–	–	–	1v	2v1r	4v2r	5v4r

Note that the conjecture (9.23) manifests its truth in these tables for $g \leq 4$ by showing that in the columns of Tn sharpness occurs on or below the "reverse diagonal" $n + \sigma/2 = g+1$. The bottom entry 1v on that diagonal comes from the $(2, 2g+1)$-torus knot, as identified in Proposition 9.2.4.

Chapter 10

Outlook

The final chapter concludes with a less formal and more abbreviated review of some further research related to generators, which is otherwise out of the scope of this book. This should help seeing the work here as a part of a larger setting of applications of Theorem 2.11.8 and its relation to concepts we had no opportunity to discuss.

10.1 Legendrian invariants and braid index

A *contact structure* on a smooth 3-manifold is a 1-form α with $\alpha \wedge d\alpha \neq 0$ (which is equivalent to the non-integrability of the plane distribution defined by $\ker\alpha$). In the following, we consider the 1-form $\alpha = dx + y\,dz$ on $\mathbb{R}^3(x,y,z)$, which is called the *standard contact space*. A *Legendrian knot* is a smooth embedding $\mathcal{K} : S^1 \to \mathbb{R}^3$ with $\alpha\left(\frac{d\mathcal{K}}{dt}\right) \equiv 0$. Each such knot has its underlying topological knot type $K = [\mathcal{K}]$ and two fundamental invariants in contact geometry known as the *Thurston–Bennequin number* or *Thurston–Bennequin invariant* $tb(\mathcal{K})$ and *Maslov index* $\mu(\mathcal{K})$. See [FT, Fe] for an excellent introductory account on this subject.

Definition 10.1.1 The Thurston–Bennequin number $tb(\mathcal{K})$ of a Legendrian knot $\mathcal{K}$ in the standard contact space is the linking number of $\mathcal{K}$ with $\mathcal{K}'$, where $\mathcal{K}'$ is obtained from $\mathcal{K}$ by a push-forward along a vector field transverse to the (hyperplanes of the) contact structure.

The Maslov (rotation) index $\mu(\mathcal{K})$ of $\mathcal{K}$ is the degree of the map

$$t \in S^1 \mapsto \frac{\mathrm{pr}\,\frac{d\mathcal{K}}{dt}(t)}{\left|\mathrm{pr}\,\frac{d\mathcal{K}}{dt}(t)\right|} \in S^1\,,$$

where pr $: \mathbb{R}^3 \to \mathbb{R}^2 \simeq \mathbb{C}$ is the projection $(x,y,z) \mapsto (y,z)$.

Both invariants $tb(\mathcal{K})$ and $\mu(\mathcal{K})$ can be interpreted in terms of a regular diagram of the (topological) knot $[\mathcal{K}]$, and thus it was recently realized that the theory of polynomial invariants of knots and links in S^3 can be applied in the context of Legendrian knots to give inequalities for tb and μ. In particular, we have a Legendrian version of the left inequality of Morton and Franks–Williams (2.25),

$$tb(\mathcal{K}) + |\mu(\mathcal{K})| \leq \min\deg_l P([\mathcal{K}]) - 1\,, \tag{10.1}$$

which was translated to this context by Tabachnikov and Fuchs [FT]. See also [Fe].

On the other hand, a purely topological estimate follows from a Legendrian version of Bennequin's inequality (Theorem 2.4.2),

$$tb(\mathcal{K}) + |\mu(\mathcal{K})| \leq 2g([\mathcal{K}]) - 1\,, \tag{10.2}$$

and a later improvement of Rudolph [Ru3], who showed

$$tb(\mathcal{K}) + |\mu(\mathcal{K})| \leq 2g_s([\mathcal{K}]) - 1\,, \tag{10.3}$$

where $g_s(K)$ is the slice (4-ball) genus of K.

While the r.h.s. of (10.2) and (10.3) are invariant w.r.t. taking the mirror image, the l.h.s. are strongly sensitive, so we have

$$\tau'(K) \leq \tau(K) \leq 2g_s(K) - 1 \leq 2g(K) - 1 \tag{10.4}$$

for any topological knot type K, where

$$\tau'(K) := \max\{tb(\mathcal{K}) + |\mu(\mathcal{K})| \,:\, [\mathcal{K}] = K\}$$

and

$$\tau(K) := \max\{\,tb(\mathcal{K}) + |\mu(\mathcal{K})| \,:\, [\mathcal{K}] \in \{K, !K\}\,\} = \max(\tau'(K), \tau'(!K))\,.$$

In [K], Kanda used an original argument and the theory of convex surfaces in contact manifolds to show that the inequality $\tau' \leq 2g - 1$ can get arbitrarily unsharp, i.e., there exist knots K_i with $\tau'(K_i) - 2g(K_i) \to -\infty$. In Kanda's paper, all K_i are alternating pretzel knots, and hence of genus 1, so that for these examples in fact we also have $\tau'(K_i) - 2g_s(K_i) \to -\infty$.

In [St18], this result was extended using Theorem 2.11.8.

Theorem 10.1.2 Let $\{K_i\}$ be a sequence of knots, such that only finitely many of the K_i have the same skein polynomial. Then $2\tilde{g}(K_i) - \min(\tau'(K_i), \tau'(!K_i)) \to \infty$. If additionally $\tilde{g}(K_i) \leq C$ for some constant C, then even $\min(\tau'(K_i), \tau'(!K_i)) \to -\infty$.

This can be applied to Cromwell's [Cr] homogeneous knots (which include alternating and positive knots).

Corollary 10.1.3 Bennequin's inequality becomes arbitrarily unsharp on any sequence of (Legendrian embeddings of distinct) properly mirrored homogeneous knots.

Theorem 10.1.2 uses Theorem 2.11.8 to first deduce:

Theorem 10.1.4 The set

$$\{P_K : \text{span}_l P_K \le b, \tilde{g}(K) \le g\}$$

is finite for any natural numbers g and b.

The other application we gave of Theorem 10.1.4 is related to a result of Birman and Menasco [BM] (building again on Bennequin's work [Be]), namely, that there are only finitely many knots of given genus and given braid index. Theorem 10.1.4 implies this result for homogeneous knots. In fact, we obtain that the lower bound for the braid index coming from the MWF inequality (2.24) gets arbitrarily large for homogeneous knots of given genus.

Corollary 10.1.5 There are only finitely many homogeneous knots K of given genus and given $\text{span}_l P_K$.

10.2 Minimal genus and fibering of canonical surfaces

In [St14], we introduced a procedure of regularization (see Lemma 5.2.3) for knot generators. We used this, together with invariants derived from the Jones polynomial (degrees, congruences, and the Fiedler–Polyak–Viro Gauss diagram formulas for its Vassiliev invariants), to examine the existence of genus-minimizing diagrams for almost alternating and almost positive knots.

It is easy to see using tongue moves (§2.6) that any almost alternating link L has almost alternating diagrams D whose canonical surfaces have arbitrarily large genus (in particular, not all have minimal genus).

In [St14], we prove among others:

Theorem 10.2.1 There are almost alternating knots with minimal genus diagrams but with no almost alternating minimal genus diagrams.

There is also a corresponding statement on almost positive knots. Note that by Bennequin's inequality Theorem 2.4.2 for an almost positive diagram D of a knot K, we have $g(K) \le g(D) \le g(K) + 1$.

Theorem 10.2.2 There exist almost positive knots none respectively all of whose almost positive diagrams have minimal genus.

These results should be compared with Theorem 2.4.1.

Further, we use Theorem 2.11.8 in [St2] to prove that for given genus, the canonical surface is generically minimal.

Definition 10.2.3 If

$$\mathcal{C} \subset \mathcal{D}$$

are classes of diagrams *or* knots, we say that the inclusion (10.5) is *asymptotically dense*, if

$$\lim_{n\to\infty} \frac{\#\big(\mathcal{C}\cap\{c(D)=n\}\big)}{\#\big(\mathcal{D}\cap\{c(D)=n\}\big)} = 1\,.$$

(This is the definition for diagrams; for knots, replace "D" by "K".)

Theorem 10.2.4 ([St2]) For each $g>0$,

$$\{\,g(D)=g,\ D \text{ minimal genus}\,\} \subset \{\,g(D)=g\,\} \tag{10.5}$$

is asymptotically dense.

In [St24], we apply the description of canonical genus 2 surfaces (Theorem 2.11.11), the methods of Gabai [Ga, Ga2, Ga3] to detect minimal genus and fiber property, and some computational effort to prove two related theorems.

Theorem 10.2.5 No knot has genus $g=1$ and canonical genus $\tilde{g}=2$.

The other result of disk decomposition, proved by essentially the same method, relates to the description of Alexander polynomials of fibered knots.

Neuwirth, who introduced fibered knots, conjectured that for given Alexander polynomial only finitely many fibered knots occur. This conjecture is true for genus 1 but was completely refuted by Morton by constructing an infinite number of fibered knots for any Alexander polynomial of degree at least 2. We show in contrast that the conjecture is "almost" true in genus 2 for *canonical* fiber surfaces.

Theorem 10.2.6 Only finitely many monic degree 2 polynomials are realized as Alexander polynomial of infinitely many (genus 2) knots with canonical fiber surfaces.

This seems the first somewhat broader statement toward confirming (at least) partially Neuwirth's conjecture.

10.3 Wicks forms, markings, enumeration of alternating knots by genus

Definition 10.3.1 A *maximal Wicks form* w is a cyclic word in the free group over an alphabet with the following three conditions:

1) Each letter a and a^{-1} appears exactly once in w.
2) w contains no subwords of the form $a^{\pm 1}a^{\mp 1}$.
3) (maximality) Whenever $a^{\pm 1}b^{\pm 1} \in w$ and $b^{\pm 1}c^{\pm 1} \in w$ are subwords of w (with the signs independently choosable), we have also $c^{\pm 1}a^{\pm 1} \in w$ (for proper to be chosen signs).

Two Wicks forms w, w' are *equivalent* up to cyclic permutation and permutation of letters (and inverses).

They were first studied by Wicks, then by Comerford-Edmunds, Culler, etc. Bacher-Vdovina [BV] related Wicks forms to duals of 1-vertex triangulations of oriented surfaces, thus:

- The number of letters in a maximal Wicks form is $6g-3$ for some $g > 0$.
- Label the edges of a $12g-6$-gon X by letters of w and reverse the orientation induced from the one of X on edges corresponding to inverses of letters.
- Identify the edges labeled by each letter and its inverse according to their orientation.
- The resulting surface S is orientable of genus g. Call g the *genus* of the Wicks form.
- The boundary ∂X becomes after identification a *3-valent (cubic)* graph $G \subset S$, which is the 1-skeleton of 1-face cell complex (edges of $X \simeq$ letters $a^{\pm 1}$, vertices $\simeq$ triples in maximality property). Thus, its dual is a 1-vertex triangulation.

Definition 10.3.2 [SV] We say a Wicks form w is *maximal planar* if its graph $G \subset S$ planar and *3-connected.*

The reason for attention to this property lies in a relation to maximal generators.

Lemma 10.3.3

$$\left\{ \begin{array}{c} \text{maximal genus } g \\ \text{knot generators} \end{array} \right\} \simeq \left\{ \begin{array}{c} \text{maximal planar Wicks} \\ \text{forms of genus } g \end{array} \right\}. \tag{10.6}$$

We will try to explain this relationship.

First, note that by Whitney's theorem, when G 3-connected, it has a unique planar embedding.

Let us say that $G \subset S$ is *cellular* if $S \setminus G$ consists of disks. This embedding can be identified in the planar embedding of G by designating which side of S is visible near each vertex of G when the three incident edges are ordered cyclically as in the planar embedding.

This gives a $+/-$ *marking*

$$O : V(G) \to \{+,-\} \tag{10.7}$$

of the vertices of G in planar embedding. Each marking gives a *thickening* by the rule

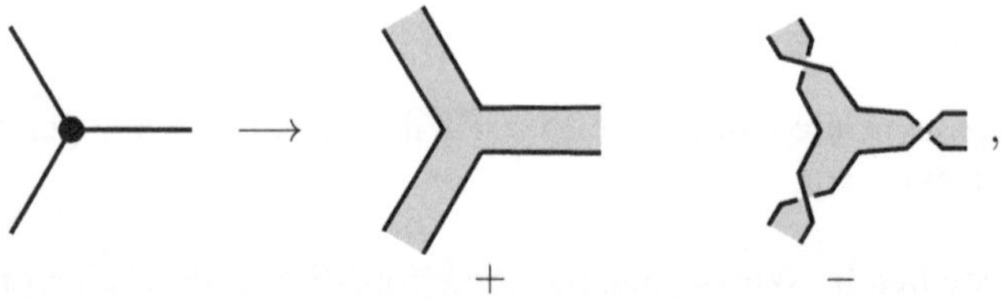

(10.8)

from which one reconstructs $S \supset G$ by gluing disks into the boundary components.

Definition 10.3.4 For a marking (10.7) and thickening (10.8) we call $S = S_{G,O}$.

The boundary $L_{G,O} = \partial S_{G,O}$ is a link, coming with a particular diagram, call it $D_{G,O}$, and it is assumed with the orientation induced from $S_{G,O}$. Then let $n_{G,O} = n(L_{G,O})$ be the number of boundary components.

Given G and O, one obtains a maximal link generator diagram $D'_{G,O}$ from $D_{G,O}$ by adding two crossings on edges between two $+$ marked vertices and switching crossings so that $D'_{G,O}$ becomes alternating. Note that the Seifert graph $\Gamma(D'_{G,O})$ is a (bipartite) reduced bisection of G, and $n(D'_{G,O}) = n_{G,O}$.

The marking O of G (or thickening $S_{G,O}$) corresponds to a (maximal planar) Wicks form iff $n_{G,O} = 1$, i.e., O is a *knot marking*. (The Wicks form thus depends also on O, not only on G.) We can see (10.6) better by starting with a knot marking O of G on the left. The corresponding maximal generator diagram on the right of (10.6) is then $D'_{G,O}$.

Vdovina introduces 3 elementary operations that construct a Wicks form of genus $g+1$ from one of genus g. We show only their effect on the graphs G, which we call *graphic α, β and γ construction:*

α β γ

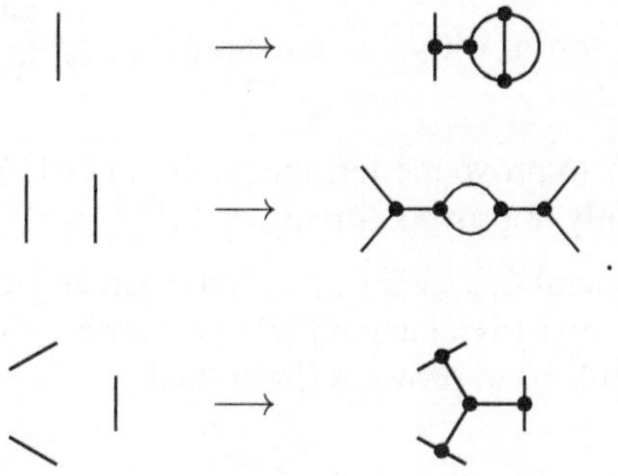

In [SV], we applied this framework to the enumeration of alternating knots by genus.

A fundamental tool is the Flyping Theorem 2.6.2. In a generator of k $\sim$-equivalence classes, there are $\binom{n}{k-1} + O(n^{k-2})$ alternating diagrams of n crossings. The Flyping theorem helps taking care of flypes and symmetries, and so we obtain Theorem 2.11.9. This also leads to (2.44).

We further see that the leading contribution for $a_{n,g}$ comes from maximal generators. Note that (also by Theorem 3.1.1) maximal knot generators of genus g have $6g-3$ $\sim$-equivalence classes.

Let us call p_g in Theorem 2.11.9 the *period*. We already wrote that p_g is in general huge (something like $\mathrm{lcm}(1,\dots,6g-3)$), but the leading terms $[P_{g,n}]_{\max}$ of $P_{g,n}$ have only period 2. We identified these terms in [SV]. To account for parity, let a maximal generator be *even/odd* depending on its crossing number. Let then

$$C_{g,\text{even/odd}} = \#\{ \text{ maximal genus } g \text{ even/odd generators } \}.$$

Theorem 10.3.5 [SV]

$$[P_{g,\text{even/odd}}]_{\max} = \frac{n^{6g-4}}{(6g-4)!} \cdot C_{g,\text{even/odd}}\,.$$

Furthermore, we have the maximal generators are special, and as for Theorem 2.11.9 above, we obtain (with Definition 10.2.3 for knots) the following.

Proposition 10.3.6 [SV] Among alternating knots of given genus, special alternating ones are asymptotically dense.

To count $C_{g,*}$, we had to adapt Bacher-Vdovina's enumeration techniques (to maximal planar Wicks forms).

Theorem 10.3.7 [SV]

$$400 \leq \liminf_{g\to\infty} \sqrt[g]{C_{g,*}} \leq \limsup_{g\to\infty} \sqrt[g]{C_{g,*}} \leq \frac{2^{20}}{3^6} \approx 1438.38\,. \qquad (10.9)$$

We use Vdovina's γ to prove the left inequality in (10.9). Later in [St10], the right inequality was slightly improved: $\limsup \leq (2^{130/7}31^{2/7})/3^6 \approx 1425.39$.

Be aware that in general $C_{g,\mathrm{even}} \neq C_{g,\mathrm{odd}}$; thus, odd and even crossing genus g alternating knots have a non-trivial asymptotic proportion. Combinatorialists have done enough work to enable us to prove in [St10] that

$$\lim_{g\to\infty} \frac{C_{g,\mathrm{even}}}{C_{g,\mathrm{odd}}} = 1\,,$$

and this convergence is exponentially fast (in g). There is also some evidence that

$$C_{g,\mathrm{odd}} > C_{g,\mathrm{even}} \text{ when } g > 2\,,$$

but this, in turn, looks like a hopeless combinatorial problem.

10.4 Crossing numbers

In [St8], we use Theorem 3.1.1 to give some estimates on the crossing numbers of reduced semiadequate diagrams. Many of them use the Kauffman polynomial F.

Theorem 10.4.1 Let L be an oriented non-split link, which is not a $(2,k)$-torus link and D be a reduced A-semiadequate diagram of L. Then

$$c(D) \leq \frac{17a_-(L) - 9\chi(L)}{2} + \max\deg_z F(L)\,, \qquad (10.10)$$

where $a_-(L)$ is as in (2.21).

We also prove there (unrelated to generators):

Theorem 10.4.2 If D is a reduced positive diagram of a non-split link L, then $c(D) \leq c(L) + \chi(L)$.

In [St2], we proved some partial analogue of Theorem 2.8.5 for positive diagrams: Every positive genus 2 knot has a positive minimal (crossing) diagram. In [St11], we proved an opposite statement for genus 3: There are infinitely many positive knots with no positive minimal diagrams.

We used Theorem 10.4.2 to extend Proposition 10.3.6 to positive knots:

Proposition 10.4.3 [St8] The inclusion

$$\{\,K \text{ special alternating}, g(K)=g\,\} \subset \{\,K \text{ positive}, g(K)=g\,\}$$

is asymptotically dense.

A consequence of this and Proposition 2.10.1 is that for positive knots of a given genus, the signature is asymptotically maximal:

$$\lim_{n\to\infty} \frac{1}{|\mathcal{P}_{n,g}|} \sum_{K\in\mathcal{P}_{n,g}} \sigma(K) \;=\; 2g\,,$$

where

$$\mathcal{P}_{n,g} := \left\{\, K : K \text{ positive},\;\; g(K)=g,\, c(K)=n \,\right\}.$$

It provides one main piece of support for the conjecture that the signature of a positive knot has an increasing lower bound in terms of the genus; see for this [St15]. The result in Chapter 6 also fits with this conjecture, which seems very difficult to resolve in the general case.

10.5 Canonical genus bounds hyperbolic volume

By the work of W. Thurston, most knots (and links) L are *hyperbolic*, that is, their complement $S^3\setminus L = H^3/\Gamma$ is the quotient of hyperbolic 3-space H^3 by some (properly discontinuously acting) group Γ of isometries. Moreover, the volume of the quotient is finite and is called the *volume of* L, $\mathrm{vol}\,(L)$.

Let us meet the convention that

$$\mathrm{vol}\,(L) := 0 \text{ when } L \text{ not hyperbolic.} \tag{10.11}$$

The context in which Brittenham [Br] discovered Theorem 2.11.8 was his following observation.

Theorem 10.5.1 (Brittenham)

$$\sup\{\,\mathrm{vol}\,(K)\,:\,\tilde g(K)=g\,\} < \infty. \tag{10.12}$$

Let $g_f(K)$ be the *free genus*, the minimal genus of free surfaces S (i.e., such for which $S^3\setminus S$ a handlebody). Canonical surfaces are free, so

$$g(K) \le g_f(K) \le \tilde g(K)\,.$$

These are often (though not always) equalities. In particular, see Theorem 2.4.1 and (10.5). The finiteness (10.12) does not apply more generally:

$$\sup\{\,\mathrm{vol}\,(K)\,:\,g_f(K)=g\,\}=\infty,$$

as proved by Brittenham [Br2] for $g=1$ and by myself for $g\geq 2$ [St26]. But of course, it does apply for alternating knots (because of Theorem 2.4.1), and the below arguments will easily show that the supremum is the same as in (10.12) (for arbitrary knots).

Let us briefly review Brittenham's argument.

Proof of Theorem 10.5.1. By Thurston's hyperbolic surgery theorem,

$$\sup\ \mathrm{vol}\,(\mathrm{series})=\mathrm{vol}\,(\textit{limit link})$$

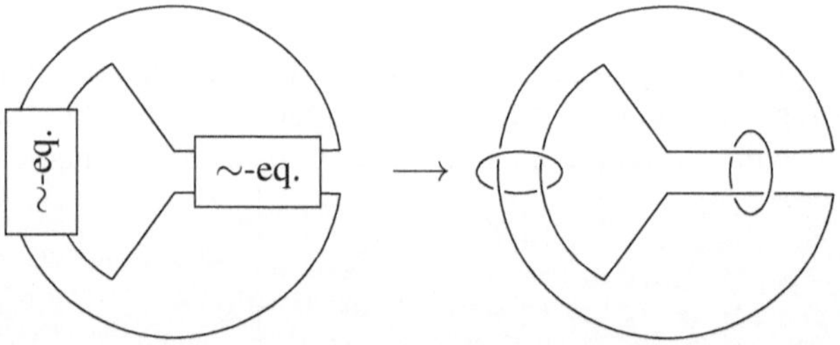

(10.13)

Because there are only finitely many series, there are only finitely many limit links, and thus the supremum (10.12) is the maximal volume of such a link. □

It is also possible to discuss links in the context of (10.12), in which case one better replaces (canonical) genus by (canonical) Euler characteristic. Let us assume again the links bound no disconnected Seifert surfaces S. This is not a significant restriction, for only disks and annuli in S can augment the number of components arbitrarily for bounded below χ, but such links are not hyperbolic. For alternating links, we saw in §2.8.2 that they bound disconnected Seifert surfaces only if they are split, and such links, again, are not hyperbolic.

Set

$$v_\chi=\sup\{\,\mathrm{vol}\,(L)\,:\,\tilde\chi(L)=\chi\,\}. \qquad (10.14)$$

We can also additionally fix the number of components: If $\tilde\chi(L)=\chi$, then L has $n=2-\chi,\chi,\dots,\chi \bmod 2(\in\{1,2\})$ components. Let

$$v_{n,\chi}=\sup\{\,\mathrm{vol}\,(L)\,:\,\tilde\chi(L)=\chi,\,n(L)=n\,\}. \qquad (10.15)$$

Theorem 10.5.2 ([St10] and [St25]). We have

$$v_{n,\chi}=v_\chi\,,$$

i.e., the maximal volume is independent of the number of components.

This theorem is an outcome of the study of the structure behind the limit links in the proof of Theorem 10.5.1.

First, it is easy to see from Hirasawa's algorithm, as well as the treatment of special generators in [SV] (most of which was recaptured in Chapters 3 and 4) that maximal (generator) series suffice to realize the maximal volume. These limit links can be associated as L_G to a 3-connected 3-valent planar graph G by

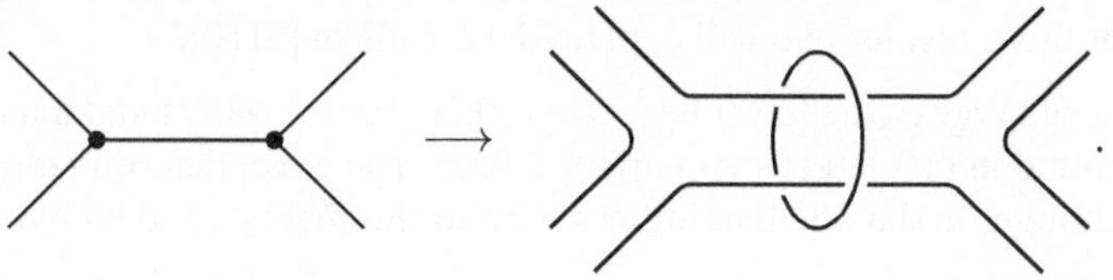
(10.16)

Then

$$v_\chi := \max_{\chi(G)=\chi} \operatorname{vol}(L_G), \tag{10.17}$$

the maximum taken over 3-connected 3-valent planar graphs. For a given number of components, the version of (10.17) connects to markings (§10.3):

$$v_{n,\chi} := \max_{\chi(G)=\chi} \operatorname{vol}(L_G), \tag{10.18}$$

changes by taking only G with n-component markings.

Theorem 10.5.2 is proved by showing that the maximal volume in (10.18) is realized by L_G for G with an n-component marking. This study is connected to the *graph volume* $\operatorname{vol}(G)$ defined by van der Veen [vV] and to operations on graphs he uses that control this volume. It will go too far for us to discuss graph volume here. Let us just mention that (from Andreev's theorem) one can ascertain planar 3-connected G to be hyperbolic, and there is a direct relationship

$$\operatorname{vol}(L_G) = \operatorname{vol}(G) + 2V_8 \cdot v(G),$$

where v is the number of vertices, and

$$V_8 \approx 3.66386 = \operatorname{vol}(\text{regular ideal octahedron}).$$

By using an estimate of Atkinson on polyhedral volume we have then

Proposition 10.5.3

$$V_8(-6\chi - 2) \leq v_\chi \leq V_8(-7\chi - 1). \tag{10.19}$$

More precisely, we will see that there is a "stable volume-χ ratio"

$$\delta = \lim_{\chi \to -\infty} \frac{v_\chi}{(-2-6\chi)V_8} = \sup_{\chi<0} \frac{v_\chi}{(-2-6\chi)V_8},$$

and

$$1 \leq \delta \leq \frac{7}{6}. \tag{10.20}$$

One can also compute many values of v_χ. For knots, we were able to go up until genus 10, where $v_{-19} \approx 444.7966230$, and for genus 11 we have at least a heuristical value, $v_{-21} \approx 493.7021266(?)$. It is clear that we cannot determine all generators or even all maximal generators in such ranges; thus some ideas were needed to upgrade the computer for these results. We will explain the details in [St10].

The computation also raises the lower bound in (10.20) to ≈ 1.08394 and thus shows that the upper bound in (10.19) is sharp up to $< 10\%$. The exact determination of v_χ remains open, though, and a challenging problem in the theory of ideal hyperbolic polyhedra.

Note also that by virtue of (10.11), this whole discussion completely bypasses non-hyperbolic links, and it does not say anything about which of the links in (10.14) are hyperbolic and which not. In general, the description of (non-)hyperbolic links among a meaningful class is a non-trivial problem. In this regard, we have (to appear at a separate place) the following application of Theorem 2.11.11 and the work on surfaces outlined in §10.2:

Theorem 10.5.4 If K is a non-trivial knot of canonical genus at most 2, then K is non-hyperbolic if and only if it is 3_1, 5_1, or the connected sum of two knots of canonical genus 1.

This also confirms the first non-trivial case of the following obvious conjecture (which is easy to prove for the genus and also known for the free genus by work of Ozawa [O2]):

Conjecture 10.5.5 The canonical genus of knots is additive under connected sum:

$$\tilde{g}(K_1 \# K_2) = \tilde{g}(K_1) + \tilde{g}(K_2).$$

10.6 The relation between volume and the sl_N polynomial

In the theory of Vassiliev invariants (following Kontsevich and Bar-Natan [BN3]), one can associate to a (semisimple) Lie algebra $\mathfrak{g}$ an invariant, *weight system*, $W_\mathfrak{g}$ of (uni-)trivalent graphs G.

For $\mathfrak{g} = sl_N$, the weight system $W_{sl_N}(G) = W_N(G)$ is a polynomial in N for a fixed trivalent graph G. That is, W_N is a polynomial invariant of trivalent graphs. This sl_N

polynomial is known to be connected to thickenings. With the definitions of §10.3, $W_N(G)$ can be calculated by:

$$W_N(G) = W_{N,+}(G) - W_{N,-}(G),$$

with

$$W_{N,+/-}(G) = \sum_{O \text{ even/odd}} N^{n_{G,O}}.$$

It is easy to observe

$$2N(N^2-1) \mid W_N(G),$$

and that $W_{N,\pm}(G)$ have only even/odd powers depending on $\chi(G)$. Moreover, $\max\deg W_{N,\pm}(G) \le 2-\chi(G)$. Call O *spherical* if $n_{G,O} = 2-\chi(G)$, which is the same as saying that the thickening $S_{G,O}$ is a planar surface.

The central piece of attention given to the polynomial W_N is likely the remarkable observation of Bar-Natan [BN2] that it carries the Four-Color Theorem (4CT) in itself.

Theorem 10.6.1 [Bar-Natan's version of 4CT]

$$\begin{array}{ccc} \#\{\text{ spherical markings of } G\,\} \neq 0 & \Rightarrow & \#\{\text{ four-colorings of } G\,\} \neq 0 \\ \| & & \wr\| \ (\text{if } G \text{ planar}) \\ [W_N(G)]_{2-\chi} \quad \neq 0 & \Rightarrow & W_2(G) \quad \neq 0 \end{array}$$

Little else known on $W_N(G)$, though. Bacher-Vdovina's work provides one further non-trivial piece of information.

Proposition 10.6.2 (see [St10]) $[W_N(G)]_1 = 0$ for all (not necessarily 3-connected or planar) 3-valent graphs G of (odd) $\chi < -1$.

We already came across (counting) markings in relation to the enumeration of alternating knots by genus. Thus, this procedure is very close to a calculation of W_N. (An important difference is that in enumeration, automorphisms of G must be taken into account.) This suggests why we calculated W_N for a large number of 3-valent (planar 3-connected) graphs. These computations raised the following question. See [St10] for a longer discussion; a partial but to-be-published reference is [St25].

Question 10.6.3 Is there a relation between $W_N(G)$ and $\mathrm{vol}(G)$ for a planar 3-connected 3-valent graph G?

Even the more direct question, whether one determines the other, creates concern. It is relatively easy to find two graphs with equal volume but different polynomial W_N (examples start at $\chi = -9$).

However, the reverse implication, does $W_N(G)$ determine $\mathrm{vol}(G)$, turned out to be a startling challenge to refute. There seems no *a priori* reason for such a relationship, but it was only after 3–4 months of computation and skillful programming until graphs $G_{1,2}$ emerged with $\mathrm{vol}(G_1) \neq \mathrm{vol}(G_2)$ and $W_N(G_1) = W_N(G_2)$. The simplest examples have $\chi = -15$.

One should not assume that the scarcity of such examples is owed to the lack of coincidences of W_N. We considered mainly *cyclically 4-connected (c4c)* graphs (for reasons too far removed to discuss here), that is, graphs whose only 3-cuts are stars of vertices. For example, for $\chi = -12$ there are 1357 c4c graphs, with 617 coincidences of W_N (i.e., W_N takes 740 different values). For these and tens of thousands more coincidences (we computed up to $\chi = -17$ and partially $\chi = -18$), the volume also coincides. Even if it differs, the relative error is stunningly small: In all observed instances,

$$\frac{\left|\,\mathrm{vol}(G_1) - \mathrm{vol}(G_2)\,\right|}{\mathrm{vol}(G_1)} < 6 \cdot 10^{-7}.$$

This veils Question 10.6.3 in quite a mystery and hints to a possible link between the Vassiliev invariant polynomial, and volume of ideal polyhedra in hyperbolic space, whose study could be worthwhile.

10.7 Everywhere equivalent links

A link diagram is said to be *everywhere equivalent* if all diagrams obtained by switching exactly one crossing represent the same link. For knot diagrams, if this knot is the unknot, we say the diagram is *everywhere trivial.* It is a folklore conjecture that the trefoil and figure-8-knot are the only non-trivial knots with everywhere trivial diagrams, and there is a candidate list of 6 of their diagrams (up to 8 crossings). It is conjectured that this list is complete, and let us call this statement the Everywhere Trivial Diagram Conjecture.

In [St27], we used Theorems 2.11.11 and 4.1 to verify that everywhere equivalent knot diagrams of low genus conform to a conjecturally exhaustive family of such diagrams given in [St27]. As a special case, we have a partial confirmation of the Everywhere Trivial Diagram Conjecture.

Bibliography

[Ad] C. C. Adams, Das Knotenbuch, Spektrum Akademischer Verlag, Berlin, 1995 (The knot book, W. H. Freeman & Co., New York, 1994).

[Ad2] —— " ——, Toroidally alternating knots and links, *Topology* **33(2)** (1994), 353–369.

[Ad3] —— " —— et al., Almost alternating links, *Topol. Appl.* **46** (1992), 151–165.

[Al] J. W. Alexander, Topological invariants of knots and links, *Trans. Amer. Math. Soc.* **30** (1928), 275–306.

[Al2] —— " ——, A lemma on systems of knotted curves, *Proc. Nat. Acad. Sci. U.S.A.* **9** (1923), 93–95.

[AB] —— " —— and G. B. Briggs, On types of knotted curves, *Ann. Math.* **28** (1927), 562–586.

[An] C. N. Aneziris, The mystery of knots. Computer programming for knot tabulation, *Series on Knots and Everything* **20**, World Scientific Publishing Co., Inc., River Edge, NJ, 1999. ISBN: 981-02-3878-9

[Ar] E. Artin, Theory of braids, *Ann. of Math.* **48** (1947), 101–126.

[Av] D. Avis, `lrs`, program package of reverse search algorithm for vertex enumeration/convex hull problems, `http://cgm.cs.mcgill.ca/~avis/C/lrs.html`.

[BMo] Y. Bae and H. R. Morton, The spread and extreme terms of Jones polynomials, *J. Knot Theory Ramifications* **12(3)** (2003), 359–373.

[BV] R. Bacher and A. Vdovina, Counting 1-vertex triangulations of oriented surfaces, *Discrete Mathematics* **246(1-3)**, 13–27.

[BN] D. Bar-Natan, Knot Atlas, online knot invariant tables and calculation software, `http://www.math.toronto.edu/~drorbn/KAtlas/index.html`.

[BN2] —— " ——, Lie algebras and the four color theorem, *Combinatorica* **17(1)** (1997), 43–52.

[BN3] —— " ——, On the Vassiliev knot invariants, *Topology* **34** (1995), 423–472.

[BN4] —— " ——, On Khovanov's categorification of the Jones polynomial, *Algebraic and Geometric Topology* **2** (2002), 337–370.

[Be] D. Bennequin, Entrelacements et équations de Pfaff, *Soc. Math. de France, Astérisque* **107-108** (1983), 87–161.

[Bi] J. S. Birman, New points of view in knot theory, *Bull. Amer. Math. Soc.* **28(2)** (1993), 253–287.

[Bi2] ——— " ———, Braids, links and mapping class groups, *Ann. of Math. Studies* **82**, Princeton, 1976.

[BM] ——— " ——— and W. W. Menasco, Studying links via closed braids VI. A non-finiteness theorem, *Pacific J. Math.* **156** (1992), 265–285.

[BM2] ——— " ——— and ——— " ———, Studying links via closed braids V. The unlink, *Trans. Amer. Math. Soc.* **329(2)** (1992), 585–606.

[BS] F. Bonahon and L. C. Siebenmann, New geometric splittings of classical knots and the classification and symmetries of arborescent knots, monograph available on `http://www-rcf.usc.edu/~fbonahon/Research/Preprints/Preprints.html`

[BLM] R. D. Brandt, W. B. R. Lickorish and K. Millett, A polynomial invariant for unoriented knots and links, *Inv. Math.* **74** (1986), 563–573.

[BMc] G. Brinkmann and B. McKay, `plantri`, a program for generation of certain types of planar graphs, `http://cs.anu.edu.au/~bdm/plantri/`.

[Br] M. Brittenham, Bounding canonical genus bounds volume, preprint (1998), available at `http://www.math.unl.edu/~mbritten/personal/pprdescr.html`.

[Br2] ——— " ———, Free genus one knots with large volume, *Pacific J. Math.* **201(1)** (2001), 61–82.

[Bu] G. Burde, Das Alexanderpolynom der Knoten mit zwei Brücken, *Arch. Math. (Basel)* **44(2)** (1985), 180–189.

[BZ] ——— " ——— and H. Zieschang, Knots, de Gruyter, Berlin, 1986.

[Ca] A. Caudron, Classification des noeuds et des enlacements, Prepublication Math. d'Orsay, Orsay, France, Université Paris-Sud (1980).

[CDM] S. Chmutov, S. Duzhin and J. Mostovoy, Introduction to Vassiliev knot invariants, Cambridge University Press, Cambridge, 2012. ISBN: 978-1-107-02083-2

[Ch] D. A. Chalcraft, On the braid index of links with nested diagrams, *Math. Proc. Cambridge Philos. Soc.* **111(2)** (1992), 273–281.

[CG] T. D. Cochran and R. E. Gompf, Applications of Donaldson's theorems to classical knot concordance, homology 3-spheres and Property P, *Topology* **27(4)** (1988), 495–512.

[Co] J. H. Conway, On enumeration of knots and links, in "Computational Problems in abstract algebra" (J. Leech, ed.), 329–358. Pergamon Press, 1969.

[Cr] P. R. Cromwell, Homogeneous links, *J. London Math. Soc. (series 2)* **39** (1989), 535–552.

[Cr2] ——— " ———, Knots and links, Cambridge University Press 2004.

[CM] ——— " ——— and H. R. Morton, Positivity of knot polynomials on positive links, *J. Knot Theory Ramif.* **1** (1992), 203–206.

[CM2] ——— " ——— and ——— " ———, Distinguishing mutants by knot polynomials, *Jour. of Knot Theory and its Ramifications* **5(2)** (1996), 225–238.

[Cw] R. Crowell, Genus of alternating link types, *Ann. of Math.* **69(2)** (1959), 258–275.

[Cw2] ——— " ———, Nonalternating links, *Illinois J. Math.* **3** (1959), 101–120.

[DL] O. Dasbach and X.-S. Lin, A volume-ish theorem for the Jones polynomial of alternating knots, math.GT/0403448, *Pacific J. Math.* **231(2)** (2007), 279–291.

[DL2] —— ” —— and —— ” ——, On the head and the tail of the colored Jones polynomial, *Compositio Math.* **142(5)** (2006), 1332–1342.

[DH] M. Dehn and P. Heegaard, Analysis situs, *Enzyk. der Math. Wiss.* **III AB 3**, Leipzig (1907), 153–220.

[DS] E. Denne and J. Sullivan, The distortion of a knotted curve, *Proc. Amer. Math. Soc.* **137(3)** (2009), 1139–1148.

[DT] C. H. Dowker and M. B. Thistlethwaite, Classification of knot projections, *Topol. Appl.* **16** (1983), 19–31.

[E] R. Elwes, The revenge of the Perko pair, web document http://richardelwes.co.uk/2013/08/14/the-revenge-of-the-perko-pair/.

[Fe] E. Ferrand, On Legendrian knots and polynomial invariants, *Proc. Amer. Math. Soc.* **130(4)** (2002), 1169–1176.

[FW] J. Franks and R. F. Williams, Braids and the Jones–Conway polynomial, *Trans. Amer. Math. Soc.* **303** (1987), 97–108.

[F&] P. Freyd, J. Hoste, W. B. R. Lickorish, K. Millett, A. Ocneanu and D. Yetter, A new polynomial invariant of knots and links, *Bull. Amer. Math. Soc.* **12** (1985), 239–246.

[FT] D. Fuchs and S. Tabachnikov, Invariants of Legendrian and transverse knots in the standard contact space, *Topology* **36(5)** (1997), 1025–1053.

[Fu] K. Fukuda, cdd and cddplus, software for calculation of convex polytopes, http://www.ifor.math.ethz.ch/~fukuda/cdd_home/cdd.html.

[FR] R. Furmaniak and S. Rankin, Knotilus, a knot visualization interface, http://srankin.math.uwo.ca/.

[FG] D. Futer and F. Guéritaud, Angled decompositions of arborescent link complements, math/0610775, *Proc. London Math. Soc.* **98(2)** (2009), 325–364.

[Ga] D. Gabai, Foliations and genera of links, *Topology* **23** (1984), 381–394.

[Ga2] —— ” ——, The Murasugi sum is a natural geometric operation, Low-dimensional topology (San Francisco, Calif., 1981), *Contemp. Math.* **20**, 131–143, Amer. Math. Soc., Providence, RI, 1983.

[Ga3] —— ” ——, The Murasugi sum is a natural geometric operation II, Combinatorial methods in topology and algebraic geometry (Rochester, N.Y., 1982), 93–100, *Contemp. Math.* **44**, Amer. Math. Soc., Providence, RI, 1985.

[Gr] S. Garoufalidis, Does the Jones polynomial determine the signature of a knot?, preprint math.GT/0310203.

[GHY] H. Goda, M. Hirasawa and R. Yamamoto, Almost alternating diagrams and fibered links in S^3, *Proc. London Math. Soc.* **83(2)** (2001), 472–492.

[Ha] R. I. Hartley, On two-bridged knot polynomials, *J. Austral. Math. Soc. Ser. A* **28(2)** (1979), 241–249.

[HL] J. Hass and J. C. Lagarias, The number of Reidemeister moves needed for unknotting, *J. Am. Math. Soc.* **14(2)** (2001), 399–428.

[HN] —— " —— and T. Nowik, Unknot diagrams requiring a quadratic number of Reidemeister moves to untangle, `http://arxiv.org/abs/0711.2350`, *Discrete & Computational Geometry* **44(1)** (2010), 91–95.

[Hy] C. Hayashi, A lower bound for the number of Reidemeister moves for unknotting, *J. Knot Theory Ramifications* **15(3)** (2006), 313–325.

[Hi] M. Hirasawa, Triviality and splittability of special almost alternating links via canonical Seifert surfaces, *Topology Appl.* **102(1)** (2000), 89–100.

[Hi2] —— " ——, The flat genus of links, *Kobe J. Math.* **12(2)** (1995), 155–159.

[HS] —— " —— and A. Stoimenow, Examples of knots without minimal string Bennequin surfaces, *Asian J. Math.* **7(3)** (2003), 435–446.

[Ho] C. F. Ho, A polynomial invariant for knots and links – Preliminary report, *Abstracts Amer. Math. Soc.* **6** (1985), 300.

[H] J. Hoste, The enumeration and classification of knots and links, Handbook of Knot Theory (W. Menasco and M. Thistlethwaite, eds.), Elsevier (2005), 209–232.

[HT] —— " —— and M. Thistlethwaite, KnotScape, a knot invariant calculation and table access program, available at `http://www.math.utk.edu/~morwen`.

[HTW] —— " ——, —— " —— and J. Weeks, The first 1,701,936 knots, *Math. Intell.* **20 (4)** (1998), 33–48.

[J] V. F. R. Jones, A polynomial invariant of knots and links via von Neumann algebras, *Bull. Amer. Math. Soc.* **12** (1985), 103–111.

[J2] —— " ——, Hecke algebra representations of braid groups and link polynomials, *Ann. of Math.* **126** (1987), 335–388.

[Jn] I. D. Jong, Alexander polynomials of alternating knots of genus two, *Osaka J. Math.* **46(2)** (2009), 353–371.

[Jn2] —— " ——, Alexander polynomials of alternating knots of genus two II, *J. Knot Theory Ramifications* **19(8)** (2010), 1075–1092.

[K] Y. Kanda, On the Thurston-Bennequin invariant of Legendrian knots and nonexactness of Bennequin's inequality, *Invent. Math.* **133(2)** (1998), 227–242.

[Kf] L. H. Kauffman, An invariant of regular isotopy, *Trans. Amer. Math. Soc.* **318** (1990), 417–471.

[Kf2] —— " ——, State models and the Jones polynomial, *Topology* **26** (1987), 395–407.

[Kf3] —— " ——, Formal knot theory, *Mathematical Notes* **30**, Princeton University Press, Princeton, NJ, 1983.

[Kf4] —— " ——, Knots and physics (second edition), World Scientific, Singapore 1993.

[La] M. Lackenby, The volume of hyperbolic alternating link complements, with an appendix by I. Agol and D. Thurston, *Proc. London Math. Soc.* **88(1)** (2004), 204–224.

[L] W. B. R. Lickorish, An introduction to knot theory, *GTM* **175**, Springer-Verlag, New York, 1997.

[LM] —— " —— and K. C. Millett, A polynomial invariant for oriented links, *Topology* **26 (1)** (1987), 107–141.

[LT] ——— " ——— and M. B. Thistlethwaite, Some links with non-trivial polynomials and their crossing numbers, *Comment. Math. Helv.* **63** (1988), 527–539.

[Li] C. N. Little, Non-alternate $\pm$ knots, *Proc. Roy. Soc. Edinburgh* **39(3)** (1900), 771–778, plates I, II, III.

[MT] J. Malesic and P. Traczyk, Seifert circles, braid index and the algebraic crossing number, *Topology Appl.* **153(2-3)** (2005), 303–317.

[Me] W. W. Menasco, Closed incompressible surfaces in alternating knot and link complements, *Topology* **23 (1)** (1986), 37–44.

[MT] ——— " ——— and M. B. Thistlethwaite, The Tait flyping conjecture, *Bull. Amer. Math. Soc.* **25 (2)** (1991), 403–412.

[Mr] M. Morishita, On certain analogies between knots and primes, *J. Reine Angew. Math.* **550** (2002), 141–167.

[Mo] H. R. Morton, Seifert circles and knot polynomials, *Proc. Camb. Phil. Soc.* **99** (1986), 107–109.

[MR] ——— " ——— and H. J. Ryder, Mutants and $\mathrm{SU}(3)_q$ invariants, The Epstein birthday schrift, *Geom. Topol. Monogr.* **1**, Geom. Topol. Publ., Coventry, 1998, 365–381.

[Mu] K. Murasugi, Jones polynomial and classical conjectures in knot theory, *Topology* **26** (1987), 187–194.

[Mu2] ——— " ———, On the genus of the alternating knot, *J. Math. Soc. Japan* **10** (1958), 94–105, 235–248.

[Mu3] ——— " ———, On a certain subgroup of the group of an alternating link, *Amer. J. Math.* **85** (1963), 544–550.

[Mu4] ——— " ———, On the braid index of alternating links, *Trans. Amer. Math. Soc.* **326 (1)** (1991), 237–260.

[Mu5] ——— " ———, On a certain numerical invariant of link types, *Trans. Amer. Math. Soc.* **117** (1965), 387–422.

[Mu6] ——— " ———, On the Alexander polynomial of alternating algebraic knots, *J. Austral. Math. Soc. Ser. A* **39(3)** (1985), 317–333.

[MP] ——— " ——— and J. Przytycki, An index of a graph with applications to knot theory, *Mem. Amer. Math. Soc.* **106** (1993).

[MP2] ——— " ——— and ——— " ———, The skein polynomial of a planar star product of two links, *Math. Proc. Cambridge Philos. Soc.* **106(2)** (1989), 273–276.

[MS] ——— " ——— and A. Stoimenow, The Alexander polynomial of planar even valence graphs, *Adv. Appl. Math.* **31(2)** (2003), 440–462.

[NO] S. Negami and K. Okita, The splittability and triviality of 3-bridge links, *Trans. Am. Math. Soc.* **289** (1985), 253–280.

[Oh] Y. Ohyama, On the minimal crossing number and the braid index of links, *Canad. J. Math.* **45(1)** (1993), 117–131.

[O] M. Ozawa, Closed incompressible surfaces in complements of positive knots, *Comment. Math. Helv.* **77** (2002), 235–243.

[O2] ——— " ———, Additivity of free genus of knots, *Topology* **40(4)** (2001), 659–665.

[OS] P. S. Ozsváth and Z. Szabó, Holomorphic disks and knot invariants, *Adv. Math.* **186(1)** (2004), 58–116.

[Pe] K. A. Perko Jr., On the classification of knots, *Proc. Amer. Math. Soc.* **45** (1974), 262–266.

[Pe2] —— " ——, Invariants of 11 crossing knots, Publications Math. d'Orsay, 1980.

[Pe3] —— " ——, How to see the linking branch curves in covering spaces of colored knots, unpublished note.

[Pe4] —— " ——, Primality of certain knots, *Topology Proc.* **7** (1982), 109–118.

[PT] J. Przytycki and K. Taniyama, Almost positive links have negative signature, *Abstr. AMS* **12(3)** (1991), 327; `http://arxiv.org/abs/0904.4130`, *J. Knot Theory Ramifications* **19(2)** (2010), 187–289.

[PTr] —— " —— and P. Traczyk, Conway algebras and skein equivalence of links, *Proc. Amer. Math. Soc.* **100** (1987), 744–748.

[QW] C. V. Quach Hongler and C. Weber, On the topological invariance of Murasugi special components of an alternating link, *Math. Proc. Cambridge Philos. Soc.* **137(1)** (2004), 95–108.

[RS] Q. I. Rahman and G. Schmeisser, Analytic theory of polynomials, London Mathematical Society Monographs, New Series **26**, Oxford University Press, 2002.

[RFS] S. Rankin, O. Flint and J. Schermann, Enumerating the prime alternating knots. I., *J. Knot Theory Ramifications* **13(1)** (2004), 57–100.

[Re] K. Reidemeister, Knotentheorie, Springer-Verlag, Berlin, 1932.

[Re2] —— " ——, Das exakte Denken der Griechen: Beitrge zur Deutung von Euklid, Plato, Aristoteles, Claassen und Goverts, Hamburg, 1949.

[Ro] D. Rolfsen, Knots and links, Publish or Perish, 1976.

[Ru] L. Rudolph, Positive links are strongly quasipositive, *Geometry and Topology Monographs* **2** (1999), Proceedings of the Kirbyfest, 555–562. See also `http://www.maths.warwick.ac.uk/gt/GTMon2/paper25.abs.html`.

[Ru2] —— " ——, Braided surfaces and Seifert ribbons for closed braids, *Comment. Math. Helv.* **58** (1983), 1–37.

[Ru3] —— " ——, Quasipositivity as an obstruction to sliceness, *Bull. Amer. Math. Soc. (N.S.)* **29(1)** (1993), 51–59.

[Sc] A. Schrijver, Theory of linear and integer programming, Wiley-Interscience Series in Discrete Mathematics, John Wiley & Sons, Ltd., Chichester, 1986.

[Se] H. Seifert, Über das Geschlecht von Knoten, *Math. Ann.* **110** (1934), 571–592.

[S] D. S. Silver, Knot theory's odd origins, *American Scientist* **94(2)** (2006), 158–165.

[SSW] —— " ——, A. Stoimenow and S. G. Williams, Euclidean Mahler measure and Twisted Links, *Algebr. Geom. Topol.* **6** (2006), 581–602.

[St] A. Stoimenow, Positive knots, closed braids, and the Jones polynomial, `math/9805078`, *Ann. Scuola Norm. Sup. Pisa Cl. Sci.* **2(2)** (2003), 237–285.

[St2] —— " ——, Knots of (canonical) genus two, `math.GT/0303012`, *Fund. Math.* **200(1)** (2008), 1–67.

[St3] ——— " ———, Gauss sums on almost positive knots, math.GT/9803073, *Compositio Math.* **140(1)** (2004), 228–254.

[St4] ——— " ———, Knots of genus one, *Proc. Amer. Math. Soc.* **129(7)** (2001), 2141–2156.

[St5] ——— " ———, On polynomials and surfaces of variously positive links, math.GT/0202226, *Jour. Europ. Math. Soc.* **7(4)** (2005), 477–509.

[St6] ——— " ———, On some restrictions to the values of the Jones polynomial, *Indiana Univ. Math. J.* **54 (2)** (2005), 557–574.

[St7] ——— " ———, The signature of 2-almost positive knots, *J. Of Knot Theory and Its Ram.* **9(6)** (2000), 813–845.

[St8] ——— " ———, On the crossing number of semiadequate links, *Forum Math.* **26(4)** (2014), 1187–1246.

[St9] ——— " ———, Gauss sum invariants, Vassiliev invariants and braiding sequences, *Jour. of Knot Theory and Its Ramifications* **9(2)** (2000), 221–269.

[St10] ——— " ———, On dual triangulations of surfaces, Lie algebras, and alternating link volumes, preprint.

[St11] ——— " ———, Coefficients and non-triviality of the Jones polynomial, *J. Reine Angew. Math. (Crelle's J.)* **657** (2011), 1–55; see also math.GT/0606255.

[St12] ——— " ———, Genus generators and the positivity of the signature, *Alg. Geom. Top.* **6** (2006), 2351–2393.

[St13] ——— " ———, Properties of closed 3-braids, preprint math.GT/0606435.

[St14] ——— " ———, Minimal genus and fibering of canonical surfaces, preprint.

[St15] ——— " ———, Bennequin's inequality and the positivity of the signature, *Trans. Amer. Math. Soc.* **360(10)** (2008), 5173–5199.

[St16] ——— " ———, Newton-like polynomials of links, *Enseign. Math. (2)* **51(3-4)** (2005), 211–230.

[St17] ——— " ———, Square numbers, spanning trees and invariants of achiral knots, math.GT/0003172, *Comm. Anal. Geom.* **13(3)** (2005), 591–631.

[St18] ——— " ———, A property of the skein polynomial with an application to contact geometry, math.GT/0008126, *J. Differential Geom.* **77(3)** (2007), 555–566.

[St19] ——— " ———, On the number of links and link polynomials, *Quart. J. Math. Oxford* **55(1)** (2004), 87–98.

[St20] ——— " ———, On the definition of graph index, *J. Austral. Math. Soc.* **94(3)** (2013), 417–429.

[St21] ——— " ———, Application of braiding sequences III: Concordance of positive knots, *Internat. J. Math.* **26(7)** (2015) 1550050 1–36, DOI: 10.1142/S0129167X15500500

[St22] ——— " ———, Tait's conjectures and odd crossing number amphicheiral knots, *Bull. Amer. Math. Soc.* **45** (2008), 285–291, arXiv:0704.1941.

[St23] ——— " ———, Knot data tables, web document http://www.stoimenov.net/stoimeno/homepage/ptab/index.html.

[St24] ——— " ———, Minimal genus and fibering of canonical surfaces via disk decomposition, *LMS J. of Comp. and Math.* **17(1)** (2014), 77–108.

[St25] —— " ——, A theorem on graph embedding with a relation to hyperbolic volume, accepted in *Combinatorica*.

[St26] —— " ——, Realizing Alexander polynomials by hyperbolic links, *Expo. Math.* **28(2)** (2010), 133–178.

[St27] —— " ——, Everywhere equivalent and everywhere different knot diagrams, *Asian J. Math.* **17(1)** (2013), 95–137.

[StK] —— " ——, The crossing number and maximal bridge length of a knot diagram, with an appendix by M. Kidwell, *Pacific J. Math.* **210(1)** (2003), 189–199.

[STV] —— " ——, V. Tchernov and A. Vdovina, The canonical genus of a classical and virtual knot, *Geometriae Dedicata* **95(1)** (2002), 215–225.

[SV] —— " —— and A. Vdovina, Counting alternating knots by genus, *Math. Ann.* **333** (2005), 1–27.

[ST] C. Sundberg and M. B. Thistlethwaite, The rate of growth of the number of prime alternating links and tangles, *Pacific Journal of Math.* **182 (2)** (1998), 329–358.

[T] P. G. Tait, On knots II, *Trans. Roy. Soc. Edinburgh* **32** (1884), 327–340, plate 44.

[Ta] K. Taniyama, A partial order of knots, *Tokyo J. Math.* **12(1)** (1989), 205–229.

[Th] M. B. Thistlethwaite, On the Kauffman polynomial of an adequate link, *Invent. Math.* **93(2)** (1988), 285–296.

[Th2] —— " ——, On the algebraic part of an alternating link, *Pacific J. Math.* **151(2)** (1991), 317–333.

[Th3] —— " ——, On the structure and scarcity of alternating links and tangles, *J. Knot Theory Ramifications* **7(7)** (1998), 981–1004.

[Th4] —— " ——, A spanning tree expansion of the Jones polynomial, *Topology* **26(3)** (1987), 297–309.

[Tr] P. Traczyk, Non-trivial negative links have positive signature, *Manuscripta Math.* **61** (1988), 279–284.

[Tr2] —— " ——, Reducing braid index by Murasugi–Przytycki operation, *J. of Knot Theory and Its Ram.* **20(1)** (2011), 223–229.

[Ts] T. Tsukamoto, The almost alternating diagrams of the trivial knot, `math/0605018`, *J. Topology* **2** (2009), 77–104.

[V] A. A. Vdovina, Constructing orientable Wicks forms and estimation of their number, *Communications in Algebra* **23(9)** (1995), 3205–3222.

[vV] R. van der Veen, The volume conjecture for augmented knotted trivalent graphs, *Algebr. Geom. Topol.* **9(2)** (2009), 691–722.

[Vo] P. Vogel, Representation of links by braids: A new algorithm, *Comment. Math. Helv.* **65** (1990), 104–113.

[Wo] S. Wolfram, Mathematica: A system for doing mathematics by computer, Addison-Wesley, 1989.

[Ya] S. Yamada, The minimal number of Seifert circles equals the braid index, *Invent. Math.* **88** (1987), 347–356.

[Yo] Y. Yokota, Polynomial invariants of positive links, *Topology* **31(4)** (1992), 805–811.

[Zu] L. Zulli, The rank of the trip matrix of a positive knot diagram, *J. Knot Theory Ramif.* **6 (2)** (1997), 299–301.

Glossary

∼-equivalence is an equivalence relation of crossings of a link diagram related to generators.

alternating is a property of a link diagram to have all crossing underpasses followed by an overpass and vice versa, and of links having such diagrams.

amphicheiral or **achiral** is the property of a knot to be isotopic to its mirror image.

Bennequin surface is a minimal genus Seifert surface of a link spanned on a braid.

bipartite a property of a graph to have only edges between a subdivision of its vertices.

bisection an operation on graphs adding a vertex of valence 2, and a graph obtained from another graph by a sequence of such operations.

braid a collection of strings carrying a group structure and giving a link under closure operation.

bridge a piece of the strand of a link diagram containing only crossing overpasses; **tunnel** is (here) its mirror image.

canonical refers to Seifert surfaces obtained by his method, and to the genera of such surfaces and of the link diagrams carrying these surfaces.

checkerboard can refer to a coloring of the regions of a link diagram, and to a graph constructed from that coloring.

chiral is a knot which is not amphicheiral.

closure can refer to

1. (topologically) the set of all points obtainable from within a given set as a limit
2. **braid closure** a link obtained from a braid by identifying strand endpoints

component of a link is one of the embedded circles of that link.

composite is a link or link diagram which is a connected sum (i.e., not prime).

connected sum is a binary operation on links and link diagrams.

contraction an operation on graphs shrinking an edge or the star of a vertex.

convex hull the closure of all convex linear combinations of a set.

crossing is a transverse self-intersection of a link diagram with under/over information.

diagram a planar curve representation of knots and links with crossing information.

flype one basic type of diagram move, for alternating diagrams interrelating all these diagrams of a given link.

generator can refer to

1. **group generator** one of a set of elements whose words (with inverses) exhaust a group
2. **genus generator** (most commonly here) a link or link diagram in a set from which one can obtain all diagrams of given genus under crossing changes, flypes and $\bar{t}_2'$ moves.

genus a basic topological quantity classifying closed compact orientable surfaces and associated to links and link diagrams in relation to their Seifert surfaces.

Hirasawa move a link diagram move eliminating separating Seifert circles (and ultimately making the diagram special) and preserving the diagram genus.

index can refer to

1. **braid index** the minimal number of strands of a braid representation of a link
2. **index of a Seifert circle** the number of changes between crossings attached from in- and outside
3. **graph index** one of several sorts of recursively defined quantities for (bipartite) graphs, used for braid index estimates.

invariant a quantity assigned to some geometric structure (here, most commonly, link diagram), which does not change under deformations of that structure (here, most commonly, Reidemeister moves).

Kauffman bracket a version of the Jones polynomial defined using crossing splicings.

knot a smooth embedding of a circle in $\mathbb{R}^3$ or S^3.

knot table aims to list (prime, mirrored) knots of given crossing number exactly once.

link a smooth embedding of a collection of circles in $\mathbb{R}^3$ or S^3.

link polynomial an invariant of links taking values in some ring of (Laurent) polynomials.

Murasugi sum or **$*$-product** is a binary operation on link diagrams identifying a Seifert circle (and more generally on links and Seifert surfaces).

mutation is an operation on links and link diagrams not altering many invariants (incl. link polynomials and hyperbolic volume).

MWF inequality gives a lower bound for the braid index in terms of the skein polynomial.

negative is the mirror image of positive (crossing, diagram, or link).

polytope closure of the intersection of a finite number of halfspaces.

positive is a sign of a crossing in a link diagram, a property of a link diagram to have only such crossings, and a link to have such a diagram.

prime is a link or link diagram which is not a connected sum.

reduced can refer to

1. a link diagram with no nugatory crossings
2. a bisection of a graph with no adjacent vertices of valence of 2.

region a connected component of the planar complement of a link diagram.

Reidemeister moves three fundamental moves interrelating all diagrams of a given link.

Seifert circle a planar loop obtained from a link diagram after smoothing out crossings.

Seifert surface is a compact oriented surface bounding a link.

series is (here) a set of diagrams obtained from a generator by flypes, crossing changes and $\bar{t}'_2$ moves.

signature an integer valued link invariant.

skein relation a relation between the values of an invariant on three links with diagrams differing locally.

slice knot a knot bounding a (smooth) disk in B^4.

smoothing is a local operation on link diagrams eliminating a crossing.

special a property of a link diagram to have no separating Seifert circles; mostly used in combination with alternating, and then also for the links with such diagrams.

split a link diagram with disconnected planar curves, or a link with such diagram.

split union a binary operation on links and link diagrams adding split parts.

star is the set of edges incident from a vertex in a graph.

state in the Kauffman bracket refers to a choice of A/B-splicings for all crossings of a link diagram.

$\bar{t}'_2$ move is a move preserving diagram genus and used to obtain all diagrams in a series of a generator.

unbisection is the reverse of bisection.

wave move is a link diagram move consisting of rerouting and possibly shrinking a bridge or tunnel.

writhe the sum of signs of all crossings of a link diagram.

Index

The boldfaced entries refer to the page of definition (or most detailed introduction). Some notions are used so frequently that it rather makes sense to record only the most important occurrences.